Lecture Notes in Computer Science   10731

Commenced Publication in 1973
Founding and Former Series Editors:
Gerhard Goos, Juris Hartmanis, and Jan van Leeuwen

Editorial Board

More information about this series at http://www.springer.com/series/7409

Christos Doulkeridis · George A. Vouros
Qiang Qu · Shuhui Wang (Eds.)

Mobility Analytics for Spatio-Temporal and Social Data

First International Workshop, MATES 2017
Munich, Germany, September 1, 2017
Revised Selected Papers

 Springer

Editors
Christos Doulkeridis 🆔
University of Piraeus
Piraeus
Greece

George A. Vouros 🆔
University of Piraeus
Piraeus
Greece

Qiang Qu 🆔
Shenzhen Institutes of Advanced
 Technology
Shenzhen
China

Shuhui Wang
Institute of Computing Technology
Beijing
China

ISSN 0302-9743   ISSN 1611-3349   (electronic)
Lecture Notes in Computer Science
ISBN 978-3-319-73520-7   ISBN 978-3-319-73521-4   (eBook)
https://doi.org/10.1007/978-3-319-73521-4

Library of Congress Control Number: 2017962899

LNCS Sublibrary: SL3 – Information Systems and Applications, incl. Internet/Web, and HCI

Printed on acid-free paper

This Springer imprint is published by Springer Nature
The registered company is Springer International Publishing AG
The registered company address is: Gewerbestrasse 11, 6330 Cham, Switzerland

Preface

This proceedings volume contains revised versions of the papers presented at the First International Workshop on Mobility Analytics for Spatiotemporal and Social Data (MATES 2017), held in conjunction with the 43rd International Conference on Very Large Data Bases (VLDB 2017), in Munich, Germany, on September 1, 2017.

Mobility analytics is a timely topic owing to the ever-increasing number of diverse, real-life applications, ranging from social media to land, sea, and air surveillance systems, which produce massive amounts of streaming spatiotemporal data, whose acquisition, cleaning, representation, aggregation, processing, and analysis pose new challenges for the data management community. The aim of MATES is to bring together researchers and practitioners interested in developing data-intensive applications that analyze big spatiotemporal/societal data, in order to foster the exchange of new ideas on multidisciplinary real-world problems, propose innovative solutions, and stimulate further research in the area of big spatiotemporal/societal data management and analysis. The workshop intends to bridge the gap between researchers and domain experts, most importantly to raise awareness of real-world problems in critical domains which require novel data management solutions, tailored to addressing the specific needs of each domain.

The peer-review process put great emphasis on ensuring a high quality of accepted contributions. Every paper was reviewed by at least three Program Committee (PC) members. The MATES PC accepted six submissions (46%) as full papers and another two submissions (16%) as short papers out of a total of 13 submissions. After careful revision of accepted papers, based both on comments of reviewers and discussions during the workshop, the chairs decided to allocate the same maximum page length for all papers included in this volume.

Apart from the peer-reviewed papers that were presented at the workshop, the program included two keynote speeches, one from academia and one from the industrial sector. The first keynote — "Effective and Efficient Community Search" — was given by Dr. Reynold Cheng, Associate Professor of the Department of Computer Science at the University of Hong Kong. The second keynote — "How Data Analytics Enables Advanced AIS Applications" — was given by Ernest Batty, Technical Director of IMIS Global Limited. After the workshop, both keynote speakers were invited to submit a paper describing the research objectives and future challenges in relation with their talks, and these papers are included in this volume.

This volume is structured as follows: In the first part, we include the invited papers from the keynote speakers. Then, the research papers are grouped in thematic areas: The second part concerns "Social Networks Analytics and Applications," while the third part addresses "Spatiotemporal Mobility Analytics." In this way, the grouping of research papers reflects the two major foci of the workshop, namely, mobility analytics for social data and mobility analytics for spatiotemporal data.

The editors wish to thank the PC members for helping MATES put together a program of high-quality papers that provides an up-to-date overview of the area of mobility analytics for spatiotemporal and social data. In addition, the editors would like to thank all authors for submitting their work to MATES.

On a final note, we wish to mention that this workshop was partially supported by the European Union's Horizon 2020 research and innovation programme datAcron: Big Data Analytics for Time Critical Mobility Forecasting, under grant agreement number 687591, the CAS Pioneer Hundred Talents Program, and the MOE Key Laboratory of Machine Perception at Peking University under grant number K-2017-02.

November 2017

Christos Doulkeridis
George Vouros
Qiang Qu
Shuhui Wang

Organization

Program Committee

Natalia Andrienko	Fraunhofer Institute IAIS, Germany
Alexander Artikis	NCSR Demokritos, Greece
Elena Camossi	NATO Centre for Maritime Research and Experimentation (CMRE), Italy
Christophe Claramunt	Naval Academy Research Institute, France
Jose Manuel Cordero Garcia	CRIDA, Spain
Christos Doulkeridis	University of Piraeus, Greece
Georg Fuchs	Fraunhofer Institute IAIS, Germany
Maria Halkidi	University of Piraeus, Greece
Anne-Laure Jousselme	NATO Centre for Martime Research and Experimentation (CMRE), Italy
Sofia Karagiorgou	University of Piraeus, Greece
Jooyoung Lee	Syracuse University, USA
Jiehuan Luo	Jinan University, China
Michael Mock	Fraunhofer Institute IAIS, Germany
Mohamed Mokbel	University of Minnesota, USA
Kjetil Noervaag	Norwegian University of Science and Technology, Norway
Kostas Patroumpas	University of Piraeus, Greece
Nikos Pelekis	University of Piraeus, Greece
Jiang Qingshan	Chinese Academy of Sciences, China
Qiang Qu	Shenzhen Institutes of Advanced Technology, China
Cyril Ray	Naval Academy Research Institute, France
Giorgos Santipantakis	University of Piraeus, Greece
David Scarlatti	Boeing Research and Technology Europe, Spain
Liu Siyuan	The Pennsylvania State University, USA
Yannis Theodoridis	University of Piraeus, Greece
Goce Trajcevski	Northwestern University, USA
Akrivi Vlachou	University of Piraeus, Greece
George Vouros	University of Piraeus, Greece
Shuhui Wang	Chinese Academy of Sciences, China
Raymong Wong	The Hong Kong University of Science and Technology, SAR China

Contents

On Attributed Community Search

Yixiang Fang$^{(\boxtimes)}$ and Reynold Cheng

Department of Computer Science, The University of Hong Kong,
Hong Kong, China
{yxfang,ckcheng}@cs.hku.hk

Abstract. Communities, which are prevalent in *attributed graphs* (e.g., social networks and knowledge bases) can be used in emerging applications such as product advertisement and setting up of social events. Given a graph G and a vertex $q \in G$, the *community search* (CS) query returns a subgraph of G that contains vertices related to q. In this article, we study CS over two common attributed graphs, where (1) vertices are associated with keywords; and (2) vertices are augmented with locations. For keyword-based attributed graphs, we investigate the *keyword-based attributed community* (or KAC) query, which returns a KAC for a query vertex. A KAC satisfies both *structure cohesiveness* (i.e., its vertices are tightly connected) and *keyword cohesiveness* (i.e., its vertices share common keywords). For spatial-based attributed graphs, we aim to find the *spatial-aware community* (or SAC), whose vertices are close structurally and spatially, for a query vertex in an online manner. To enable efficient KAC search and SAC search, we propose efficient query algorithms. We also perform experimental evaluation on large real datasets, and the results show that our methods achieve higher effectiveness than the state-of-the-art community retrieval algorithms. Moreover, our solutions are faster than baseline approaches. In addition, we develop the *C-Explorer* system to assist users in extracting, visualizing, and analyzing KACs.

1 Introduction

Due to the developments of gigantic social networks (e.g., Flickr, and Facebook), the topic of *attributed graphs* has attracted attention from both industry and research areas [18–20,29,37,49,53]. Essentially, an attributed graph is a graph, in which vertices or edges are associated with attributes. The attribute of a vertex often refers to its features, including its interest, hobbies, and locations, while the attribute of an edge often indicates the relationships between two vertices. In this article, we consider two typical kinds of attributed graphs, which are *keyword-based* attributed graphs, and *spatial-based* attributed graphs, and study the problem of community search on these attributed graphs.

Let us see some examples of attributed graphs. Figure 1 illustrates an attributed graph, where each vertex represents a social network user, and each edge represent the friendship between two users. The keywords of each user describe the interest of that user. Figure 2 depicts a spatial-based attributed graph with nine users in three cities, and each user has a location. The solid

© Springer International Publishing AG 2018
C. Doulkeridis et al. (Eds.): MATES 2017, LNCS 10731, pp. 1–21, 2018.
https://doi.org/10.1007/978-3-319-73521-4_1

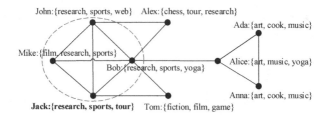

Fig. 1. A keyword-based attributed graph and an attributed community.

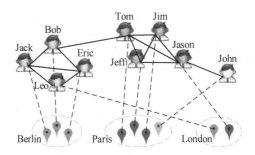

Fig. 2. A spatial-based attributed graph.

lines represent their social relationship, and the dashed lines denote their locations.

The problems related to retrieving communities from a graph can generally be classified into *community detection* (CD) and *community search* (CS). In general, CD methods aim to discovery all communities of a graph [22,38,41,42,44,49,50, 54]. These solutions are not "query-based", i.e., they are not customized for a query request (e.g., a user-specified query vertex). Besides, it is not clear how these algorithms can efficiently return a community that contain a given vertex q. Moreover, they can take a long time to find all the communities for a large graph, and so they are not suitable for quick or *online* retrieval of communities.

To solve issues above, CS methods [1,8,9,32,33,48] have been recently developed. The CS solutions aim to search the community of a specific query vertex in an "online" manner, which implies that these approaches are query-based. However, existing CS algorithms assume *non-attributed* graphs, and only use the graph structure information to find communities. Thus, it is desirable to develop methods for searching communities with the consideration of attributes. As we will show later, the use of attribute information can significantly improve the effectiveness of the communities retrieved.

In this article, we systematically study the motivation, applications, features, technical challenges, algorithms, and experimental evaluation of searching communities over these two kinds of attributed graphs. In the following, we will detail how to tackle these issues.

1.1   CS on Keyword-Based Attributed Graphs

Given an attributed graph G and a vertex $q \in G$, the *keyword-based attributed community* (or KAC) query returns one or more subgraphs of G known as *keyword-based attributed communities* (or KACs). A KAC is a kind of *community*, which consists of vertices that are closely related. Particularly, a KAC satisfies *structure cohesiveness* (i.e., its vertices are closely linked to each other) and *keyword cohesiveness* (i.e., its vertices have keywords in common). Figure 1 illustrates an AC (circled), which is a connected subgraph with vertex degree 3; its vertices {Jack, Bob, John, Mike} have two keywords (i.e., "research" and "sports") in common.

The main features of KA search are that: (1) *Ease of interpretation.* As shown in Fig. 1, a KAC contains tightly-connected vertices with similar contexts or backgrounds. Thus, a query user can focus on the common keywords or features of these vertices (e.g., the vertices of the KAC in this example contain "research" and "sports", reflecting that all members of this KAC like research and sports). (2) *Personalization.* The user of an KACs can control the semantics of the AC, by specifying a set of S of keywords. Intuitively, S decides the meaning of the AC based on the user's need. If we let $q = $ Jack, $k = 2$ and $S = \{$ "research"$\}$, the AC is formed by {Jack, Bob, John, Mike, Alex}, who are all interested in research. Thus, with the use of different keyword sets S, different "personalized" communities can be obtained. (3) *Online evaluation.* Similar to other CS solutions, we have developed efficient query algorithms for large graphs, allowing ACs to be generated quickly upon a query request.

We define a KAC based on the *minimum degree*. We formulate the keyword cohesiveness as maximizing the number of shared keywords in keyword set S. The shared keywords naturally reveal the common features among vertices (e.g., common interest of social network users). A simple way of answering a KAC query is to consider all the possible keyword combinations, and then return the subgraphs, which satisfy the minimum degree constraint and have the most shared keywords. This solution has a complexity exponential to the size of q's keyword set, so it is impractical, when q's keyword set is large.

We first propose two baseline solutions. We further develop the *CL-tree* index, which organizes the vertex keyword data in a hierarchical structure. Based on the CL-tree index, we have developed three different KAC algorithms, and they are able to achieve a superior performance. We have performed extensive experiments on large real graph datasets, and the results show that KAC query achieves higher effectiveness than existing CD and CS algorithms. Moreover, our proposed algorithms are much faster than the baseline solutions. Finally, we propose Community-Explorer (or *C-Explorer*), a web-based system that can assist users in extracting, visualizing, and analyzing communities.

Figure 3 shows the user interface of *C-Explorer* configured to run on the DBLP bibliographical network. On the left panel, a user inputs the name of an author (e.g., "jim gray") and the minimum degree of each vertex in the community she wants to have. The user can also indicate the labels or *keywords* related to her community. Once she clicks the "Search" button, the right panel will

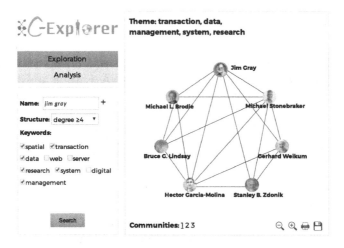

Fig. 3. Interface of *C-Explorer*.

quickly display a community of Jim Gray, which contains researchers working on database system transactions since they all share the keyword set {transaction, data, management, system, research}. *C-Explorer* implements several state-of-the-art CR algorithms, including `Global` [48], `Local` [9], and `CODICIL` [44], and provides functions for analyzing the communities. A user can also plug new CR solution into *C-Explorer* through a application programmer interface (API).

1.2 CS on Spatial-Based Attributed Graphs

Given a spatial graph G and a vertex $q \in G$, our goal is to find a subgraph of G, called a *spatial-aware community* (or SAC). An SAC is a community with high *structure cohesiveness* and *spatial cohesiveness*. The structure cohesiveness mainly measures the social connections within the community, while the spatial cohesiveness focuses on the closeness among their geo-locations. Figure 2 illustrates an SAC with three users {`Tom`, `Jeff`, `Jim`}, in which each user is linked with each other and all of them are in `Paris`.

We adopt the *minimum degree* [9,36,48] to measure the structure cohesiveness. To measure the spatial cohesiveness, we consider the *spatial circle*, which contains all the community members. In specific, given a query vertex $q \in G$, our goal is to find an SAC containing q in the smallest *minimum covering circle* (or MCC) and all the vertices of the SAC satisfy the minimum degree measure.

The main features of SAC search are that: (1) *Adaptability to location changes.* As the locations of users often change over time and their link relationship also evolve over time, SAC search can adapt to such dynamic easily, as it can answer queries in an "online" manner. Figure 4(b) shows another user's two SACs in three days, when she moves from place "C" to place "D". These real examples clearly show that a user's communities could evolve over time. (2) *Personalization.* SAC search is customized for finding communities for a

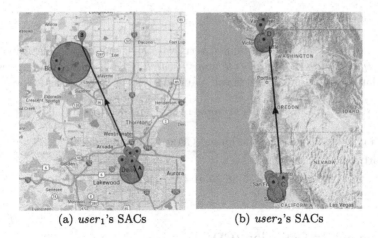

(a) $user_1$'s SACs   (b) $user_2$'s SACs

Fig. 4. SACs in Brightkite dataset.

particular query user, and the link cohesiveness of the community can also be controlled. (3) *Online evaluation.* The SAC search is able to find an SAC from a large spatial graph quickly once a query request arrives.

Since SACs achieve both high structure and spatial cohesiveness, it can be applied to many interesting applications including event recommendation (e.g., *Meetup*[1]), social marketing, and geo-social data analysis. For example, *Meetup* tracks its users' mobile phone locations, and suggests interesting location-based events to them [53]. Suppose that *Meetup* wishes to recommend an event to a user u. Then we can first find u's SAC, whose members are physically close to u. Events proposed by u's SAC member v can then be introduced to u, so that u can meet v if she is interested in v's activity. Since u's location changes constantly, u's recommendation needs to be updated accordingly.

The SAC search problem is very challenging, because the center and radius of the smallest MCC containing q are unknown. A basic exact approach takes $O(m \times n^3)$ time to answer a query, where n and m denote the numbers of vertices and edges in G. To alleviate this issue, we develop three efficient approximation algorithms with arbitrary approximation ratio, and an advanced exact algorithm, which is much faster than the basic exact algorithm. We have performed experiments on real datasets and the results show that our solutions yield better communities than those produced by existing CS and CD algorithms. Moreover, the approximation algorithms are much faster than the exact algorithms.

We organize the rest of the article as follows. We review the related works in Sect. 2. In Sect. 3, we investigate the problem of CS on keyword-based attributed graphs. In Sect. 4, we examine the problem of CS on spatial-based attributed graphs. We conclude and discuss the future work in Sect. 5.

---

[1] https://www.meetup.com/.

2   Related Work

The related works about community retrieval can generally be classified into *community detection* (CD) and *community search* (CS). Table 1 summarizes the works related to community retrieval. We review them in detail as follows.

Table 1. Classification of works in community retrieval (CR).

Graph type	Community detection (CD)	Community search (CS)
Non-attributed	[22, 25, 42]	[1, 8, 9, 32, 33, 36, 48]
Attributed	[6, 12, 26, 34, 35, 38, 39, 41, 44, 49, 50, 52, 54]	**KAC** [13, 14, 16, 17], **SAC** [15]

2.1   Community Detection (CD)

Detecting communities from a network is a fundamental research problem in network science, and it has been widely studied during the past several decades [43]. In the following, we mainly review studies about CD on attributed graphs.

CD on Keyword-Based Attributed Graphs. The clustering technique is often used to detect communities from keyword-based attributed graphs. Zhou et al. [54] considered both links and keywords of vertices to compute the vertices' pairwise similarities, and then clustered the graph based on the similarities. Ruan et al. [44] proposed a method called `CODICIL`. This solution augments the original graphs by creating new edges among vertices based on their content similarity, and then uses an effective graph sampling to boost the efficiency of clustering.

Another common approach is based on the generative models. The LDA model [4] is a classical generative statistical model, which is able to explain the observations based on some unobserved variables. In [38,41], the `Link-PLSA-LDA` and `Topic-Link LDA` models jointly model vertices' content and links based on the `LDA` model. In [49], Xu et al. developed a Bayesian probabilistic model which can capture both structures and attributes of vertices. In [45], Sachan et al. proposed to discover communities based on the topics, interaction types and the social connections among the vertices. `CESNA` [50] detects overlapping communities by assuming communities "generate" both the link and content. A discriminative approach [51], which combines the link and content analysis, has also been considered for CD on attributed graphs. However, these CD solutions are generally slow, as they often consider the pairwise distance/similarity among vertices in an entire graph. Also, they partition graphs with no reference to the query queries, and it is not clear how they can answer online queries.

CD on Spatial-Based Attributed Graphs. Many recent works identify communities from spatially constrained graphs, whose vertices are not only have links, but also associated with spatial coordinates [2]. For example, Girvan et al. [25] studied the geo-community, which is a graph of intensely connected vertices being loosely connected with others, but it is more compact in space. Guo et al. [26] proposed the average linkage (ALK) measure for clustering objects in spatially constrained graphs. In [12], Expert et al. adapted the modularity function for spatial networks and proposed a method to uncovered communities from spatial graphs. In [47], Shakarian et al. modified the well known Louvain algorithm and used a variant of Newman-Girvan modularity to mine the geographically dispersed communities from location-based social networks. In [6], Chen et al. proposed a geo-distance-based method using fast modularity maximization for identifying communities that are both highly topologically connected and spatially clustered from spatially constrained networks. We will compare our proposed methods with it in the experiments.

However, these CD algorithms are generally costly and time-consuming, as they often detect all the communities from an entire network. None of these CD methods has been shown to be able to quickly detect communities from spatial-based attributed graphs with millions or billions of vertices. Also, it is not clear how they can be adapted for online retrieval of communities from large spatial-based attributed graphs. Thus, it calls for the development of faster algorithms of performing CS on the spatial-based attributed graphs.

2.2   Community Search (CS)

To perform CS queries, people often define some measures of structure cohesiveness for a community. We classify the existing CS solutions using these measures.

Minimum degree. The minimum degree is one of the most fundamental characteristics of a graph [5,23]. In [48], Sozio et al. proposed the first CS solution, called Global. Given a graph G and a query vertex $q \in G$, Global returns the largest connected subgraph containing q as the target community in an online manner. It finds the community by iteratively removing vertices whose degrees are less than k. Cui et al. [9] also used the minimum degree measure and developed another solution Local, which uses local expansion techniques to enhance the performance of Global. In [1], Barbieri et al. further improved Global and Local and generalized the query such that it can find a community of multiple query vertices. In [36], Li et al. assumed each graph vertex has an influence value and proposed to find the top-r k-influential communities.

k-truss. The k-truss of a graph is the largest subgraph, in which each edge is contained by at least $(k - 2)$ triangles in the sub-graph [7]. In [32], Huang et al. proposed to search overlapping communities based on k-truss. In [33], Huang et al. proposed to find the closest truss communities.

Other measures. Some other classical measures (e.g., k-clique and connectivity) have also been applied to CS. In [8], Cui et al. proposed to find overlapping communities based on k-cliques. The connectivity of a graph is the minimum number of edges whose removal disconnect it [24]. In [30,31], Hu et al. proposed a community model based on the measure of connectivity.

However, these existing CS solutions generally assume non-attributed graphs, and overlook the rich information of vertices and edges that come with attributed graphs. Therefore, it is desirable to design CS algorithms for attributed graphs.

3   CS on Keyword-Based Attributed Graphs

In this section, we first formally introduce the KAC query, then present the query algorithms, and finally discuss the experimental results.

3.1   The KAC Query

We now discuss the keyword-based attributed graph model, the k-core, and the AC. We consider a keyword-based attributed graph[2] $G(V, E)$, which is undirected with vertex set V and edge set E. Each vertex $v \in V$ is associated with a set of keywords, $W(v)$. Let n and m be the corresponding sizes of V and E. The degree of a vertex v of G is denoted by $deg_G(v)$.

A community is often a subgraph that satisfies *structure cohesiveness*. In KAC query, we use the *minimum degree*, which is also used in the k-core.

Definition 1 (k-core [3,46]). *Given an integer k ($k \geq 0$), the k-core of G, denoted by H_k, is the largest subgraph of G, such that $\forall v \in H_k$, $deg_{H_k}(v) \geq k$.*

We say that H_k has an order of k. Notice that H_k may not be a connected graph [3], and its connected components, denoted by k-$\widehat{core}$s, are usually the "communities" returned by k-$\widehat{core}$ search algorithms.

Example 1. In Fig. 5(a), $\{A, B, C, D\}$ is both a 3-core and a 3-$\widehat{core}$. The 1-core has vertices $\{A, B, C, D, E, F, G, H, I\}$, and is composed of two 1-$\widehat{core}$ components: $\{A, B, C, D, E, F, G\}$ and $\{H, I\}$. The number k in each circle represents the k-$\widehat{core}$ contained in that ellipse.

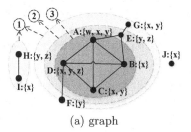

Core number	Vertices
0	J
1	F, G, H, I
2	E
3	A, B, C, D

(a) graph   (b) core number

Fig. 5. Illustrating the k-core and the KAC.

Observe that k-*core*s are "nested" [3]: given two positive integers i and j, if $i < j$, $H_j \subseteq H_i$. In Fig. 5(a), H_3 is contained in H_2, which is nested in H_1.

---

[2] Without ambiguity, all the attributed graphs mentioned in this section refer to keyword-based attributed graphs.

Definition 2 (Core number). *Given a vertex $v \in V$, its core number, denoted by $core_G[v]$, is the highest order of a k-core that contains v.*

A list of core numbers and their respective vertices for Example 1 are shown in Fig. 5(b). We now formally define the KAC query problem as follows.

Problem 1 (KAC query). Given a graph $G(V, E)$, a positive integer k, a vertex $q \in V$ and a set of keywords $S \subseteq W(q)$, return a set $\mathcal{G}$ of graphs, such that $\forall G_q \in \mathcal{G}$, the following properties hold:

- **Connectivity.** $G_q \subseteq G$ is connected and $q \in G_q$;
- **Structure cohesiveness.** $\forall v \in G_q$, $deg_{G_q}(v) \geq k$;
- **Keyword cohesiveness.** The size of $L(G_q, S)$ is maximal, where $L(G_q, S) = \cap_{v \in G_q}(W(v) \cap S)$ is the set of keywords shared in S by all vertices of G_q.

We call G_q the *keyword-based attributed community* (or KAC) of q, and $L(G_q, S)$ the *KAC-label* of G_q. In Problem 1, the first two properties are also specified by the k-$\widehat{core}$ of a given vertex q [48]. The *keyword cohesiveness* (Property 3), which is unique to Problem 1, enables the retrieval of communities whose vertices have common keywords in S. We use S to impose semantics on the KAC produced by Problem 1. By default, $S = W(q)$, which means that the KAC generated should have keywords common to those associated with q. If $S \subset W(q)$, it means that the query user is interested in forming communities that are related to some (but not all) of the keywords of q. For example, in Fig. 5(a), if $q = A$, $k = 2$ and $S = \{w, x, y\}$, the output of Problem 1 is $\{A, C, D\}$, with KAC-label $\{x, y\}$, meaning that these vertices share the keywords x and y.

We require $L(G_q, S)$ to be maximal in Property 3, because we wish the KAC(s) returned only contain(s) the most related vertices, in terms of the number of common keywords. Let us use Fig. 5(a) to explain why this is important. Using the same query ($q = A$, $k = 2$, $S = \{w, x, y\}$), without the "maximal" requirement, we can obtain communities such as $\{A, B, E\}$ (which do not share any keywords), $\{A, B, D\}$, or $\{A, B, C\}$ (which share 1 keyword).

3.2 Basic Solutions

For simplicity, we say that v contains a set S' of keywords, if $S' \subseteq W(v)$. We use $G[S']$ to denote the largest connected subgraph of G, where each vertex contains S' and $q \in G[S]$. We use $G_k[S']$ to denote the largest connected subgraph of $G[S']$, in which every vertex has degree being at least k in $G_k[S']$. We call S' a qualified keyword set for the query vertex q on the graph G, if $G_k[S']$ exists.

A simple way of answering a KAC query is to consider all the possible keyword combinations, and then return the subgraphs, which satisfy the minimum degree constraint and have the most shared keywords. This solution has a complexity exponential to the size of q's keyword set, so it is impractical, when q's keyword set is large. To alleviate this issue, we propose the following two-step framework, which is mainly based on the following *anti-monotonicity* property.

Lemma 1 (Anti-monotonicity)[3]. *Given a graph G, a vertex $q \in G$ and a set S of keywords, if there exists a subgraph $G_k[S]$, then there exists a subgraph $G_k[S']$ for any subset $S' \subseteq S$.*

The anti-monotonicity property allows us to stop examining all the super sets of $S'(S' \subseteq S)$, once have verified that $G_k[S']$ does not exist. The basic solution begins with examining the set, Ψ_1, of size-1 candidate keyword sets, *i.e.*, each candidate contains a single keyword of S. It then repeatedly executes the following two key steps, to retrieve the size-2 (size-3, ...) qualified keyword subsets until no qualified keyword sets are found.

- **Verification.** For each candidate S' in Ψ_c (initially $c=1$), mark S' as a qualified set if $G_k[S']$ exists.
- **Candidate generation.** For any two current size-c qualified keyword sets which only differ in one keyword, union them as a new expanded candidate with size-$(c+1)$, and put it into set Ψ_{c+1}, if all its subsets are qualified, by Lemma 1.

Among the above steps, the key issue is how to compute $G_k[S']$. Since $G_k[S']$ should satisfy the *structure cohesiveness* and *keyword cohesiveness*. Intuitively, we have two approaches to compute $G_k[S']$: either searching the subgraph satisfying degree constraint first, followed by further refining with keyword constraints (called `basic-g`); or vise versa (called `basic-w`).[4]

3.3   The CL-tree Index

In this section, we propose a novel index, called **CL-tree** (Core Label tree), which organizes both the k-$\widehat{cores}$ and keywords into a tree structure. Based on the index, the efficiency of answering KAC query can be improved significantly. The CL-tree index is built based on the key observation that cores are nested. Specifically, a $(k+1)$-$\widehat{core}$ must be contained in a k-$\widehat{core}$. All k-$\widehat{cores}$ can be organized into a tree structure[5].

Example 2. Consider the graph in Fig. 5(a). All the k-$\widehat{cores}$ can be organized into a tree as shown in Fig. 6(a). The height of the tree is 4. For each tree node, we attach the core number and vertex set of its corresponding k-$\widehat{core}$.

The tree structure in Fig. 6(a) can be stored compactly, as shown in Fig. 6(b). The key observation is that, for any internal node p in the tree, the vertex sets of its child nodes are the subsets of p's vertex set, because of the inclusion relationship. To save space cost, we can remove the redundant vertices that are shared by p's child nodes from p's vertex set. After such removal, we obtain a compressed tree, where each graph vertex appears only once. This structure

---

[3] All the proofs of lemmas in this article can be found in [13].

[4] All the pseudocodes of algorithms in this article can be found in [13].

[5] We use "node" to mean "CL-tree node" in Sect. 3.

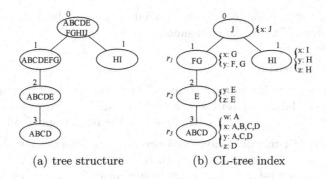

(a) tree structure   (b) CL-tree index

Fig. 6. An example CL-tree index.

constitutes the CL-tree index, the nodes of which are further augmented by inverted lists (Fig. 6(b)). The space cost of the CL-tree is linear to the size of G. To summarize, each CL-tree node p has five elements: (1) *coreNum*: the core number of the k-$\widehat{core}$; (2) *vertexSet*: a set of graph vertices; (3) *invertedList*: a list of $<key, value>$ pairs, where the *key* is a keyword contained by vertices in *vertexSet* and the *value* is the list of vertices in *vertexSet* containing *key*; and (4) *childList*: a list of child nodes;

   Using the CL-tree, the following two key operations used by our query algorithms (Sect. 3.5), can be performed efficiently.

- **Core-locating.** Given a vertex q and a core number c, find the k-$\widehat{core}$ with core number c containing q, by traversing the CL-tree.
- **Keyword-checking.** Given a k-$\widehat{core}$, find vertices which contain a given query keyword set, by intersecting the inverted lists of query keywords.

3.4   Index Construction

A simple method to build the CL-tree is build nodes recursively in a top-down manner. Specifically, we first generate the root node for 0-core, which is exactly the entire graph. Then, for each k-$\widehat{core}$ of 1-core, we generate a child node for the root node. After that, we only remain vertices with core numbers being 0 in the root node. Then for each child node, we can generate its child nodes in the similar way. This procedure is executed recursively until all the nodes are well built. We denote this index construction method by `basic`.

   Clearly, the time cost of `basic` method is $O(m \cdot k_{\max} + \widehat{l} \cdot n)$, because: (1) the k-core decomposition can be done in $O(m)$ [3]; (2) the inverted lists of each node can be built in $O(\widehat{l} \cdot n)$; and (3) in function BUILDNODE, we need to compute the connected components with a given vertex set, which costs $O(m)$ in the worst case. This may lead to low efficiency for large-scale graphs. To higher efficiency, we propose the `advanced` method, whose time and space complexities are almost linear with the size of G. The `advanced` method builds the CL-tree level by level in a bottom-up manner. Specifically, the tree nodes corresponding to larger core numbers are created prior to those with smaller core numbers. More detailed steps and analysis of `basic` and `advanced` are described in [16].

3.5   Query Algorithms

Based on the CL-tree, we propose a query algorithm, denoted by `Dec`. It first generates the candidate keyword sets, and then verifies whether them could be the shared keywords of the KACs. We illustrate the main steps as follows.

1. Generation of candidate keyword sets. `Dec` exploits the key observation that, if $S'(S' \subseteq S)$ is a qualified keyword set, then there are at least k of q's neighbors containing set S', since every vertex in $G_k[S']$ must has degree at least k. In specific, we consider q and q's neighbor vertices. For each vertex v, we only select the keywords, which are contained by S and at least k of its neighbors. Then we use these selected keywords to form an itemset, in which each item is a keyword. After this step, we obtain a list of itemsets. Then we apply the well studied frequent pattern mining algorithms to find the frequent keyword combinations, each of which is a candidate keyword set. Since our goal is to generate keyword combinations shared by at least k neighbors, we set the minimum support as k, and use the well-known FP-Growth algorithm [28].

2. Verification of candidate keyword sets. As candidates can be obtained using S and q's neighbors directly, we can verify them either incrementally, or in a decremental manner (larger candidate keyword sets first and smaller candidate keyword sets later). We choose the latter manner. The rationale behind is that, for any two keyword sets $S_1 \subseteq S_2$, the number of vertices containing S_2 is usually smaller than that of S_1, so S_2 can be verified more efficiently.

3.6   Experiments

We consider four real keyword-based attributed graphs. For each of them, each vertex has a list of neighbors as well as a set of keywords. More details of these graphs are described in [16]. To evaluate KAC queries, we set the default value of k to 6. The input keyword set S is set to the whole set of keywords contained by the query vertex. For each dataset, we randomly select 300 query vertices with core numbers of 6 or more, which ensures that there is a k-core containing each query vertex.

We have performed extensive experiments on these datasets. The detailed experimental results can be found in [16]. The general conclusions observed from the experiments are that: (1) The communities returned by KAC queries achieve higher keyword cohesiveness than the state-of-the-art CD and CS methods. For example, the Jaccard similarities of members in the KACs are higher than those of `Global` [48], `Local` [9], and CODICIL [44]. (2) The case studies on the DBLP network show that, using keywords, KAC query can find more meaningful communities than `Global` and `Local`. (3) The index-based query algorithms are over 1 to 3 orders of magnitude faster than the basic methods. For example, on the largest dataset DBpedia, a single KAC query takes less than 1 s. (4) For the index construction methods, `advanced` is much faster than `basic`.

4   CS on Spatial-Based Attributed Graphs

In this section, we first formally introduce the SAC search, then present the query algorithms, and finally discuss the experimental results.

4.1 ˙Problem Definition

Data model. We consider a spatial-based attributed graph[6] $G(V, E)$, which is an undirected graph with vertex set V and edge set E, where vertices represent entities and edges denote their relationships. For each vertex $v \in V$, it has a tuple (id, loc), where id is its ID and $loc = (x, y)$ is its spatial positions along x- and y-axis in a two-dimensional space. Let n and m be the corresponding sizes of V and E. We denote a circle with center o and radius r by $O(o, r)$, and the Euclidean distance from vertices u to v by $|u, r|$. The degree of a vertex v in a graph G is denoted by $deg_G(v)$.

Example 3. Figure 7(a) depicts a geo-social network containing 10 vertices. The solid lines linking the vertices are the edges, denoting their social relationships.

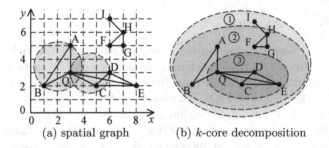

(a) spatial graph   (b) k-core decomposition

Fig. 7. An example of geo-social network.

Spatial-aware community (SAC). Conceptually, an SAC is a subgraph, G', of the graph G satisfying: (1) *Connectivity*: G' is connected; (2) *Structure cohesiveness*: all the vertices in G' are linked intensively; and (3) *Spatial cohesiveness*: all the vertices in G' are spatially close to each other.

Structure cohesiveness. We adopt the *minimum degree*, a well-accepted structure cohesiveness criterion, for measuring the structure cohesiveness of the vertices that appear in the community. Note that other criteria including k-truss [33] and k-clique [8] can also be used for SACs.

Spatial cohesiveness. To ensure high spatial cohesiveness, we require all the vertices of an SAC in a minimum covering circle (MCC) with the smallest radius. In the literature [10,11,27,40], the notion of MCC has been widely adopted to achieve high spatial compactness for a set of spatial objects.

---

[6] For simplicity, in this section we call spatial-based attributed graphs spatial graphs.

Definition 3 (MCC). *Given a set of vertices S, the MCC of S is the spatial circle, which contains all the vertices in S with the smallest radius.*

Problem 2 (SAC search). Given a graph G, a positive integer k and a vertex $q \in V$, return a subgraph $G_q \subseteq G$, and the following properties hold:

- **Connectivity.** G_q is connected and contains q;
- **Structure cohesiveness.** $\forall v \in G_q$, $deg_{G_q}(v) \geq k$;
- **Spatial cohesiveness.** The MCC of vertices in G_q satisfying Properties 1 and 2 has the minimum radius.

We call a subgraph satisfying Properties 1 and 2 a *feasible* solution, and the subgraph satisfying all the three properties the *optimal* solution (denoted by Ψ). We denote the radius of the MCC containing Ψ by r_{opt}. Essentially, SAC search finds the SAC in an MCC with the smallest radius among all the feasible solutions. In Example 3, let $C_1 = \{Q, C, D\}$ and $C_2 = \{Q, A, B\}$. The two circles in Fig. 7(a) denote the MCCs of C_1 and C_2 respectively. Let $q = Q$ and $k = 2$. The optimal solution of this query is $G[C_1]$, and $r_{opt} = 1.5$. Note that $G[C_2]$ and $G[C_1 \cup C_2]$ are feasible solutions.

4.2 SAC Search Algorithms

We first present a basic exact algorithm Exact, which takes $O(m \times n^3)$ to answer a single query. This is very time-consuming for large graphs. So we turn to design more efficient approximation algorithms. Here, the approximation ratio is defined as the ratio of the radius of MCC returned over that of the optimal solution. Inspired by the approximation algorithms, we also design a fast exact algorithm Exact+. Their approximation ratios and time complexities are summarized in Table 2, where ϵ_F and ϵ_A are parameters specified by the query user. The value $|F_1|$ is the number of "fixed vertices", which will be defined later ($|F_1| \ll n$).

AppInc is a 2-approximation algorithm, and it is much faster than Exact. Inspired by AppInc, we design another $(2 + \epsilon_F)$-approximation algorithm AppFast, where $\epsilon_F \geq 0$, which is faster than AppInc. The limitation of AppInc

Table 2. Overview of algorithms for SAC search.

Algorithm	Approximation ratio	Time complexity
Exact	1	$O(m \times n^3)$
AppInc	2	$O(mn)$
AppFast	$2 + \epsilon_F$ ($\epsilon_F \geq 0$)	If $\epsilon_F > 0$, $O(m \cdot \min\{n, \log \frac{1}{\epsilon_F}\})$ If $\epsilon_F = 0$, $O(mn)$
AppAcc	$1 + \epsilon_A$ ($0 < \epsilon_A < 1$)	$O(\frac{m}{\epsilon_A^2} \times \min\{n, \log \frac{1}{\epsilon_A}\})$
Exact+	1	$O(\frac{m}{\epsilon_A^2} \cdot \min\{n, \log \frac{1}{\epsilon_A}\} + m|F_1|^3)$

and `AppFast` is that their theoretical approximation ratios are at least 2. To achieve even lower approximation ratio, we further design another algorithm `AppAcc`, whose approximation ratio is $(1 + \epsilon_A)$, where $0 < \epsilon_A < 1$ is a value specified by the query user. Overall, these approximation algorithms guarantee that the radius of the MCC of the community has an arbitrary expected approximation ratio.

There is a trade-off between the quality of results and efficiency, i.e., algorithms with lower approximation ratios tend to have higher complexities. The pseudocodes of these algorithms can be found from [15].

The Basic Exact Algorithm. We first describe a useful lemma [11].

Lemma 2 [11]. *Given a set* S *(*$|S| \geq 2$*) of vertices, its MCC can be determined by at most three vertices in* S *which lie on the boundary of the circle. If it is determined by only two vertices, then the line segment connecting those two vertices must be a diameter of the circle. If it is determined by three vertices, then the triangle consisting of those three vertices is not obtuse.*

By Lemma 2, there are at least two or three vertices lying on the boundary of the MCC of the target SAC. We call vertices lying on the boundary of an MCC *fixed* vertices. So a straightforward method of SAC search can follow the two-step framework directly. It first finds the $k\text{-}\widehat{core}$ containing q, which is the same as `Global` does, and then returns the subgraph achieving both the structure and spatial cohesiveness by enumerating all the combinations of three vertices in the $k\text{-}\widehat{core}$. We call this method `Exact`. It completes in $O(m \times n^3)$ time.

A 2-Approximation Algorithm. In this section we present `AppInc`, which has an approximation ratio of 2. Our key observation is that, the optimal solution Ψ is very close to q. So we consider the smallest circle, denoted by $O(q, \delta)$, which is centered at q and contains a feasible solution, denoted by Φ. Let the radius of the MCC covering Φ be γ ($\gamma \leq \delta$). Note that, γ can be obtained by computing the MCC containing Φ by a linear algorithm [40]. Next, we give two lemmas:

Lemma 3. $\frac{1}{2}\delta \leq r_{opt} \leq \gamma$.

Lemma 4. *The radius of the MCC covering the feasible solution* Φ *has an approximation ratio of 2.*

`AppInc` finds Φ in an incremental manner. Specifically, it considers vertices close to q one by one incrementally, and checks whether there exists a feasible solution when a new vertex is considered. It stops once a feasible solution has been found. Clearly, `AppInc` takes $O(mn)$ time, so it is much faster than `Exact`.

A $(2 + \epsilon_F)$-Approximation Algorithm. In this section, we propose another fast approximation algorithm, called `AppFast`, which has a more flexible approximation ratio, i.e., $2 + \epsilon_F$, where ϵ_F is an arbitrary non-negative value. Instead of finding the circle $O(q, \delta)$ in an incremental manner, `AppFast` approximates the radius δ by performring binary search. We observe that, the lower and upper bounds of δ, denoted by l and u, are stated by Eq. (1):

$$l = \max_{v \in KNN(q)} |q, v|, \quad u = \max_{v \in X} |q, v|, \tag{1}$$

where X is the list of vertices of the $k\text{-}\widehat{core}$ containing q, and $KNN(q)$ contains the k nearest vertices in $X \cap nb(q)$ to q. Hence, we can approximate the radius of the circle $O(q, \delta)$ by performing binary search within $[l, u]$ until $|u - l|$ is less than a predefined small threshold α, and return an SAC denoted by Λ.

Lemma 5. *In* `AppFast`, *the radius of the MCC covering Λ has an approximation ratio of $(2 + \epsilon_F)$, if $\alpha \leq \frac{r \times \epsilon_F}{2 + \epsilon_F}$, where $\epsilon_F \geq 0$.*

A $(1 + \epsilon_A)$-Approximation Algorithm. We first present a corollary:

Corollary 1. *The center point, o, of the MCC $O(o, r_{opt})$ covering Ψ is in the circle $O(q, \gamma)$.*

Although point o is in $O(q, \gamma)$ by Corollary 1, it is still not easy to locate it exactly, since the number of its possible positions to be explored can be infinite. Instead of locating it exactly, we try to find an approximated "center", which is very close to o. In specific, we split the square containing the circle $O(q, \gamma)$ into equal-sized cells, and the size of each cell is $\beta \times \beta$ (we will explain how to set a proper value of β later). We call the center point of each cell an **anchor point**. By Corollary 1, we can conclude that o must be in one specific cell. Then we can approximate o using the anchor point of this cell, denoted by c. We consider the circle $O(c, r_{min})$, where r_{min} is the minimum radius such that it contains a feasible solution, which is denoted by Γ. We bound the value of r_{min} by Lemma 6.

Lemma 6. $r_{min} \leq r_{opt} + \frac{\sqrt{2}}{2}\beta$.

By Lemma 6, we have $\frac{r_{min}}{r_{opt}} \leq 1 + \frac{\sqrt{2}\beta}{2r_{opt}} \leq 1 + \frac{\sqrt{2}\beta}{\delta}$. Thus, we can approximate Ψ using Γ, and the approximation ratio is $(1 + \epsilon_A)$, if we let $\frac{\sqrt{2}\beta}{\delta} \leq \epsilon_A$ ($0 < \epsilon_A < 1$).

To find $O(c, r_{min})$, the basic method is that, for each anchor point p, we use `AppFast` to find the circle, which is centered at p and contains a feasible solution, and then return the minimum circle. To further improve the efficiency, we develop some optimization techniques. Specifically, we assume that all the anchor points are organized into a region quadtree [21], where the root node[7] is a square, centered at q with width 2γ. By decomposing this square into four equal-sized quadrants, we obtain its four child nodes. The child nodes of them are built in the same manner recursively, until the width of the leaf node is in $(\beta/2, \beta]$. Note that the center of each leaf node corresponds to an anchor point. To find $O(c, r_{min})$, we traverse the quadtree level by level in a top-down manner. Let

---

[7] To avoid ambiguity, we use word "node" for tree nodes in Sect. 4.

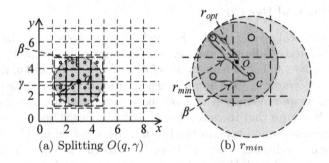

(a) Splitting $O(q, \gamma)$ (b) r_{min}

Fig. 8. Illustrating AppAcc.

r_{cur}, initialized as γ, record the smallest radius of an MCC containing a feasible solution. For each node, we first obtain the center p of its square, and then use the binary search of AppFast to approximate the smallest radius r_p, such that $O(p, r_p)$ contains a feasible solution. Based on these ideas, we develop AppAcc (Fig. 8).

Lemma 7. *If we set $\alpha' \leq \frac{1}{4}\delta\epsilon_A$ and $\beta = \frac{\delta\epsilon_A}{\sqrt{2}(2+\epsilon_A)}$ in AppAcc, where $0 < \epsilon_A < 1$ and α' is the gap between the upper bound and lower bound when stopping the binary search, the radius of the MCC covering Γ has an approximation ratio of $(1 + \epsilon_A)$.*

The Advanced Exact Algorithm. Recall that, AppAcc approximates the center, o, of the MCC covering Ψ by its nearest anchor point c, and $|o, c| \leq \frac{\sqrt{2}}{2}\beta$. Also, r_{opt} is well approximated, i.e., $\frac{r_\Gamma}{r_{opt}} \leq 1 + \epsilon_A$ where r_Γ is the radius of the MCC covering Γ. This implies that, $\frac{r_\Gamma}{1+\epsilon_A} \leq r_{opt} \leq r_\Gamma$, where $0 < \epsilon_A < 1$. So the value of r_{opt} is in a small interval, if ϵ_A is small. Besides, for any fixed vertex, f, of the MCC of Ψ, its distance to o (i.e., $|f, o|$) is exactly r_{opt}. By triangle inequality, we have

$$|f, c| \leq |f, o| + |o, c| \leq r_\Gamma + \frac{\sqrt{2}}{2}\beta, \qquad (2)$$

$$|f, c| \geq |f, o| - |o, c| \geq \frac{r_\Gamma}{1 + \epsilon_A} - \frac{\sqrt{2}}{2}\beta. \qquad (3)$$

Let us denote the rightmost items of above two inequations by r_+ and r_- respectively. Then, we conclude that, for any fixed vertex f, its distance to c is in the range $[r_-, r_+]$. If ϵ_A is very small, the gap between r_+ and r_-, i.e., $r_+ - r_- = r_\Gamma(1 - \frac{1}{1+\epsilon_A}) + \sqrt{2}\beta$, is also very small, which implies that the locations of the fixed vertices are in a very narrow annular region. Hence, a large number of vertices out of this annular region, which are not fixed vertices, can be pruned safely. Based on above analysis, we design Exact+.

4.3 Experimental Results

We consider four real spatial graphs. For each of them, each vertex has a list of graph neighbors and a 2-dimensional location. To evaluate SAC search, we set the default value of k to 4. The default values of ϵ_F and ϵ_A are set as 0.5, since these values practically result in good approximation ratios with reasonable efficiency. For each dataset, we randomly select 200 query vertices with core numbers of 4 or more, which ensures that there is a k-core containing each query vertex. More details of these graphs and experimental results are reported in [15].

The general conclusions observed from the experiments are that: (1) The communities returned by SAC search achieve higher spatial cohesiveness than the state-of-the-art CD and CS methods. For example, the radius of the MCC covering the communities returned by SAC search is much smaller than that of Global [48], Local [9], and GeoModu [6]. (2) For exact algorithms, Exact+ is over four orders of magnitude faster than Exact. For approximation algorithms, AppFast is the fastest one while AppInc is slowest one, and AppAcc is the most accurate one. (3) For moderate-size graphs, Exact+ achieves not only the highest quality results, but also reasonable efficiency. While for large graphs with millions of vertices, AppFast and AppAcc should be better choices as they are very faster.

5 Conclusions and Future Work

In this article, we investigate the problem of community search (CS) over two common attributed graphs, where (1) vertices are associated with keywords; and (2) vertices are augmented with location information. For keyword-based attributed graphs, we study the problem of keyword-based attributed community (KAC) query and find the KACs of a query vertex. Essentially, a KAC is a community that exhibits structure and keyword cohesiveness. To answer the KAC query, we develop the CL-tree index and query algorithms. Our experimental results on real datasets show that KAC queries are more effective than existing CS and CD algorithms. In addition, our solutions are faster than existing CS algorithms. For spatial-based attributed graphs, we study the spatial-aware community (SAC) search problem, which finds the community containing q within the smallest minimum covering circle (MCC). Essentially, an SAC is a community that exhibits both structure and spatial cohesiveness. We propose two exact algorithms, and three efficient approximation algorithms. The experimental results on real datasets show that, SAC search achieves better effectiveness than the existing CD and CS algorithms. Also, our algorithms are very fast. Furthermore, we develop *C-Explorer*, a system for online and interactive extracting, visualizing, and analyzing communities of a query vertex.

This article opens to a number of promising directions for the future work: (1) It would be interesting to adopt other classical metrics (e.g., k-truss and k-clique) for finding communities from attributed graphs. (2) For keyword-based attributed graphs, it would be interesting to consider other keyword cohesiveness (e.g., Jaccard similarity and string edit distance) for formulating the community models. (3) For spatial-based attributed graphs, it is of interest to examine other kinds of spatial cohesiveness measures by considering more spatial regions (e.g., squares) and the pair-wise distances of vertices. (4) For *C-Explorer*, it is worth considering more attributes of vertices and edges for searching the communities.

References

1. Barbieri, N., Bonchi, F., Galimberti, E., Gullo, F.: Efficient and effective community search. DMKD **29**(5), 1406–1433 (2015)
2. Barthélemy, M.: Spatial networks. Phys. Rep. **499**(1), 1–101 (2011)
3. Batagelj, V., Zaversnik, M.: An o(m) algorithm for cores decomposition of networks. arXiv (2003)
4. Blei, D.M., Ng, A.Y., Jordan, M.I.: Latent Dirichlet allocation. J. Mach. Learn. Res. **3**, 993–1022 (2003)
5. Bollobás, B.: The evolution of random graphs. Trans. Am. Math. Soc. **286**(1), 257–274 (1984)
6. Chen, Y., Jun, X., Minzheng, X.: Finding community structure in spatially constrained complex networks. IJGIS **29**(6), 889–911 (2015)
7. Cohen, J.: Trusses: cohesive subgraphs for social network analysis. National Security Agency Technical Report, p. 16 (2008)
8. Cui, W., Xiao, Y., Wang, H., Lu, Y., Wang, W.: Online search of overlapping communities. In: SIGMOD, pp. 277–288 (2013)
9. Cui, W., Xiao, Y., Wang, H., Wang, W.: Local search of communities in large graphs. In: SIGMOD, pp. 991–1002 (2014)
10. Elzinga, D.J., Hearn, D.W.: The minimum covering sphere problem. Manage. Sci. **19**(1), 96–104 (1972)
11. Elzinga, D.J., Hearn, D.W.: Geometrical solutions for some minimax location problems. Transp. Sci. **6**(4), 379–394 (1972)
12. Expert, P., et al.: Uncovering space-independent communities in spatial networks. PNAS **108**(19), 7663–7668 (2011)
13. Fang, Y.: Effective and efficient community search over large attributed graphs. HKU Ph.D. thesis, September 2017
14. Fang, Y., Cheng, R., Chen, Y., Luo, S., Hu, J.: Effective and efficient attributed community search. VLDB J. **26**(6), 803–828 (2017)
15. Fang, Y., Cheng, R., Li, X., Luo, S., Hu, J.: Effective community search over large spatial graphs. PVLDB **10**(6), 709–720 (2017)
16. Fang, Y., Cheng, R., Luo, S., Hu, J.: Effective community search for large attributed graphs. PVLDB **9**(12), 1233–1244 (2016)
17. Fang, Y., Cheng, R., Luo, S., Hu, J., Huang, K.: C-explorer: browsing communities in large graphs. PVLDB **10**(12), 1885–1888 (2017)
18. Fang, Y., Cheng, R., Tang, W., Maniu, S., Yang, X.: Scalable algorithms for nearest-neighbor joins on big trajectory data. TKDE **28**(3), 785–800 (2016)
19. Fang, Y., Cheng, R., Tang, W., Maniu, S., Yang, X.S.: Scalable algorithms for nearest-neighbor joins on big trajectory data. In: ICDE, pp. 1528–1529 (2016)
20. Fang, Y., Zhang, H., Ye, Y., Li, X.: Detecting hot topics from Twitter: a multiview approach. J. Inf. Sci. **40**(5), 578–593 (2014)
21. Finkel, R.A., Bentley, J.L.: Quad trees: a data structure for retrieval on composite keys. Acta Informatica **4**(1), 1–9 (1974)
22. Fortunato, S.: Community detection in graphs. Phys. Rep. **486**(3), 75–174 (2010)

23. Gaertler, M., Patrignani, M.: Dynamic analysis of the autonomous system graph. In: IPS, pp. 13–24 (2004)
24. Gibbons, A.: Algorithmic Graph Theory. Cambridge University Press, Cambridge (1985)
25. Girvan, M., Newman, M.E.J.: Community structure in social and biological networks. PNAS **99**(12), 7821–7826 (2002)
26. Guo, D.: Regionalization with dynamically constrained agglomerative clustering and partitioning (redcap). IJGIS **22**(7), 801–823 (2008)
27. Guo, T., Cao, X., Cong, G.: Efficient algorithms for answering the m-closest keywords query. In: SIGMOD, pp. 405–418. ACM (2015)
28. Han, J., Pei, J., Yin, Y.: Mining frequent patterns without candidate generation. In: SIGMOD (2000)
29. Hu, J., Cheng, R., Huang, Z., Fang, Y., Luo, S.: On embedding uncertain graphs. In: CIKM. ACM (2017)
30. Hu, J., Wu, X., Cheng, R., Luo, S., Fang, Y.: Querying minimal Steiner maximum-connected subgraphs in large graphs. In: CIKM, pp. 1241–1250 (2016)
31. Hu, J., Xiaowei, W., Cheng, R., Luo, S., Fang, Y.: On minimal steiner maximum-connected subgraph queries. TKDE **29**(11), 2455–2469 (2017)
32. Huang, X., Cheng, H., Qin, L., Tian, W., Yu, J.X.: Querying k-truss community in large and dynamic graphs. In: SIGMOD (2014)
33. Huang, X., Lakshmanan, L.V.S., Yu, J.X., Cheng, H.: Approximate closest community search in networks. PVLDB **9**(4), 276–287 (2015)
34. Kim, Y., Son, S.-W., Jeong, H.: Finding communities in directed networks. Phys. Rev. E **81**(1), 016103 (2010)
35. Leicht, E.A., Newman, M.E.J.: Community structure in directed networks. Phys. Rev. Lett. **100**(11), 118703 (2008)
36. Li, R.-H., Qin, L., Yu, J.X., Mao, R.: Influential community search in large networks. In: PVLDB (2015)
37. Li, Z., Fang, Y., Liu, Q., Cheng, J., Cheng, R., Lui, J.: Walking in the cloud: parallel simrank at scale. PVLDB **9**(1), 24–35 (2015)
38. Liu, Y., Niculescu-Mizil, A., Gryc, W.: Topic-link LDA: joint models of topic and author community. In: ICML (2009)
39. Malliaros, F.D., Vazirgiannis, M.: Clustering and community detection in directed networks: a survey. Phys. Rep. **533**(4), 95–142 (2013)
40. Megiddo, N.: Linear-time algorithms for linear programming in r3 and related problems. In: FOCS, pp. 329–338. IEEE (1982)
41. Nallapati, R.M., Ahmed, A., Xing, E.P., Cohen, W.W.: Joint latent topic models for text and citations. In: KDD (2008)
42. Newman, M.E.J., Girvan, M.: Finding and evaluating community structure in networks. Phys. Rev. E **69**(2), 026113 (2004)
43. Plantié, M., Crampes, M.: Survey on social community detection. In: Ramzan, N., van Zwol, R., Lee, J.S., Clüver, K., Hua, X.S. (eds.) Social Media Retrieval. Computer Communications and Networks, pp. 65–85. Springer, London (2013). https://doi.org/10.1007/978-1-4471-4555-4_4
44. Ruan, Y., Fuhry, D., Parthasarathy, S.: Efficient community detection in large networks using content and links. In: WWW (2013)
45. Sachan, M., et al.: Using content and interactions for discovering communities in social networks. In: WWW, pp. 331–340 (2012)
46. Seidman, S.B.: Network structure and minimum degree. Soc. Netw. **5**(3), 269–287 (1983)

47. Shakarian, P., et al.: Mining for geographically disperse communities in social networks by leveraging distance modularity. In: KDD, pp. 1402–1409 (2013)
48. Sozio, M., Gionis, A.: The community-search problem and how to plan a successful cocktail party. In: KDD (2010)
49. Xu, Z., Ke, Y., Wang, Y., Cheng, H., Cheng, J.: A model-based approach to attributed graph clustering. In: SIGMOD, pp. 505–516. ACM (2012)
50. Yang, J., McAuley, J., Leskovec, J.: Community detection in networks with node attributes. In: ICDM, pp. 1151–1156 (2013)
51. Yang, T., Jin, R., Chi, Y., Zhu, S.: Combining link and content for community detection: a discriminative approach. In: KDD (2009)
52. Yang, T., et al.: Directed network community detection: a popularity and productivity link model. In: SDM, pp. 742–753. SIAM (2010)
53. Zhang, W., et al.: Combining latent factor model with location features for event-based group recommendation. In: KDD, pp. 910–918. ACM (2013)
54. Zhou, Y., Cheng, H., Yu, J.F.: Graph clustering based on structural/attribute similarities. VLDB 2(1), 718–729 (2009)

Data Analytics Enables Advanced AIS Applications

Ernest Batty[(✉)]

IMIS Global Limited, Fareham, UK
ernie.b@imisglobal.com

Abstract. The maritime Automatic Identification System (AIS) data is obtained from many different terrestrial and satellite sources. AIS data enables safety, security, environmental protection and the economic efficiency of the maritime sector. The quality of AIS receivers is not controlled in the same manner as AIS transmitters. This has led to a situation where AIS data is not as clean as it should/could be. Added to this is the lack of accuracy and standards in entering the voyage data by the mariners such as next port of call into the AIS equipment installed on vessels. By using analytics IMIS Global Limited has been able to process the AIS data stream to eliminate a large portion of the faulty data. This has allowed the resultant AIS data to be used for more accurate detailed analysis such as the long-term vessel track, port arrival events and port departure events. New data that is derived from processing AIS data has enhanced the information available to maritime authorities enabling a significant increase in safety, security, environmental protection and economic growth. The next generation of maritime data communications technology being based AIS. This is known as the VHF Data Exchange System (VDES) and this technology now enables further opportunities. The value from the large volumes of AIS data is extracted by visual, streaming, historical and prescriptive data analytics. The datAcron project is showing the way with regards to the processing and use of AIS and resultant trajectory data.

Keywords: AIS · VDES · datAcron

1 Introduction

When AIS was conceived in the mid-1990, those working in the maritime communications industry and on the associated standards, did not fully realise the future impact that AIS would have on the maritime safety, security, economics and environmental protection. The focus was primarily on maritime safety or accident prevention with AIS defined as an 'aid to navigation'.

AIS uses a Self Organising Time Domain Multiple Access (SOTDMA) scheme in the VHF maritime band and transfers digital information from ship to shore, from ship to ship and from shore to ship, the so called 6S communication system. This uses $2 \times 2{,}250$ 26.67 ms slots to transfer data in a range of messages with standard formats.

© Springer International Publishing AG 2018
C. Doulkeridis et al. (Eds.): MATES 2017, LNCS 10731, pp. 22–35, 2018.
https://doi.org/10.1007/978-3-319-73521-4_2

AIS uses 27 standard format messages as described in the AIS standard, ITU-R M.1371-5 that include five basic message types:

1. Dynamic data
2. Static data
3. Safety Related Messages
4. Binary messages
5. System management messages

1.1 Dynamic Data

Dynamic data contains the position information of the GPS antenna connected to the AIS unit along with the Speed Over Ground (SOG), Course Over Ground (COG), Rate Of Turn (ROT) and True Heading (TH).

Along with this data, there is a Communication State parameter that adds value to understanding the AIS message and the temporal nature of the AIS dynamic messages.

1.2 Static Data

The static data concerns the ship on which the AIS unit is mounted and includes:

1. The precise location of the GPS antenna on the ship
2. The dimensions of the ship
3. The static draft of the ship
4. The name of the ship
5. The International Maritime Organisation (IMO) number of the ship (hull)
6. The call sign of the ship
7. The estimated time of arrival at the next port
8. The destination port name

Some of these data, such as the destination port, are entered by the mariner on the bridge.

1.3 Safety Related Messages

Safety Related Messages are text messages that are either broadcast and addressed to another ship or authority using the Maritime Mobile Service Identifier (MMSI) (a 9-digit identifier). The MMSI can also identify an authority or a service such as the local Coast Guard or the Maritime Rescue Coordination Centre (MRCC).

1.4 Binary Messages

Binary message can be broadcast or addressed and can contain any data to be sent between AIS units.

Binary messages are often used for telemetry or sensor data such as Internet of Things (IOT) data and includes meteorological and hydrological data.

1.5   System Management Messages

AIS system management messages are used to manage the VHF Data Link (VDL) layer load and access to it. These messages are normally under the control of the competent authority.

2   A Defining Moment for AIS

The events of the 9th of September 2001 in the United States of America added security to the maritime operations agenda and AIS was a good candidate to quickly satisfy this requirement. This need focused the global maritime sector on the identity of the vessel, its voyage history and its current location and navigational status.

AIS went from a 'nice to have' to a 'have to have' technology for all SOLAS vessels along with many work vessels within the port and off shore energy environments and vessels carrying paying passengers.

AIS receivers were installed in as many ports and waterways by various authorities. These AIS receivers were networked together and the data streamed into data processing and storage environments.

Data collected from AIS receivers started pouring into national authorities and a range of commercial terrestrial and space based services. Various commercial companies focused on getting an AIS receiver in every major port and then along most of the important waterways. The amount of AIS data increased significantly and the cost of collecting, processing and storing this data became higher than the apparent value.

The value to the users of the data was as simple as seeing ships fitted with AIS units moving on a chart. This supported Maritime Domain Awareness (MDA) initiative.

Many maritime administrations and commercial systems that were collecting AIS data started decimating data to try and reduce the amount of AIS data being streamed and stored. This reduced the amount of data being stored from one message per 5 s on average per sailing vessel to one message every few minutes.

This process continues to this day in some environments.

The decimation appeared to not have any negative impact on the services being used by the authorities and commercial sector at the time but data has been lost and some fidelity along with it.

Collecting AIS data, fusing it with any other available data and displaying this on a Common Operating Picture (COP) was the focus of many in the MDA environment.

Extracting further information from the data has now become a focus.

3   Maritime AIS Data Fidelity

Maritime data that has been decimated loses fidelity or accuracy. AIS data has spatial and temporal components. The AIS spatial components are accurate to approximately 10 m dependent on Global Navigation Satellite System (GNSS) accuracy and the availability and use of Differential GNSS. The AIS temporal components can be accurate to a single AIS slot which is 26.67 ms.

AIS data has two components:

1. TAG Block/Comment Block data/Meta Data
2. AIS message data

The TAG Block or Comment Block data is often not available in an AIS data stream or database or not considered important but this contains the source, destination and time of the message to an accuracy of one second. Depending on the system being used, additional data could be included that describes the AIS message status and quality. This has particular application in the collection and processing of satellite AIS data or so-called S-AIS data.

The TAG Block or Comment Blocks effectively doubles the amount of data that is received and processed for each AIS message transmitted from vessel. This therefore also increases the load on the data links, data processing and storage environment that use this data.

The AIS data, besides having the dynamic navigational detail of the vessel also includes some timing, slot and performance detail with regards to the AIS signal including:

1. The next slot to be transmitted in
2. The slot of the current message
3. The number of received AIS stations

Some of the AIS data is corrupted in the path between the transmitter and the Maritime Information System (MIS) that is receiving, storing, processing and disseminating the AIS data to third party applications. The performance of the MIS is described in the various international standards.

The corruption of AIS data has a number of causes which includes:

1. Substandard AIS receivers that corrupt the data
2. Substandard AIS receivers that cannot handle the load of AIS data in high traffic areas or under anomalous propagation conditions and lose data
3. Vessel Traffic Services (VTS) equipment that cannot handle the entire range of AIS messages and had to be modified to do this and corrupted AIS data in the process

There are other causes of AIS data errors and these include:

1. Incorrectly configured AIS units with primarily the type of ship, ship dimensions, GPS antenna position or MMSI being incorrect
2. Incorrectly installed AIS units which inhibits the AIS on the vessels from detecting ships in the local area and providing incorrect data
3. Mariners not updating the ETA and destination detail on a per voyage basis

With the availability of AIS data from various satellite service providers using Low Earth Orbiting (LEO) satellites, a global view was obtained of the entire maritime fleet. The increasing sophistication of the AIS receivers, associated antenna systems and the number of LEO satellites carrying AIS receivers is leading to significantly more AIS data being available to consumers of AIS data.

The issues with the security of AIS data was being highlighted by a number of entities and a widely distributed event where it was shown that it was possible to enter spoofed data into a Maritime Information System (MIS) and not have this event automatically detected but easily visible when shown on a chart.

At this same time, some of the authorities were concerned about the exact location of ships being known and this information distributed publicly and in real time. This resulted in a design and sale of various devices that could be added to ships between the external GPS and the AIS that could spoof their position by a small distance (100's of meters) but not jeopardise the AIS as an 'aid to navigation' concept.

The question facing maritime authorities and systems suppliers alike was: Could this apparent security problem with AIS technology be overcome on both the vessel and on the shore side MDA applications?

4   Data Analytics

With the rapid improvements in computer hardware performance and the growth in data storage capability of database systems and technologies, it became economical to run all AIS data through high performance complex event applications and streaming analytics applications.

The first of these was to create an application that detected faulty AIS data. Faulty AIS data could have been overtly caused (spoofed vessels) or caused by the range of reasons already noted above.

The typical functions to filter out faulty or incorrect AIS data include the following:

1. Detect incorrectly formatted messages
2. Detect message with the incorrect temporal attributes
3. Detect messages with incoherent contents
4. Detect and eliminate duplicate MMSI
5. Detect changes in static data
6. Detect changes in the data that violate the hydrodynamic performance envelope of the vessel

AIS data is now often verified against message temporal attributes, message construction and ship hydrodynamic performance which flags approximately 10% of all AIS data from some sources as having some errors. Once flagged, the severity of the issue can be determined and the message eliminated or just flagged and processed further. This data can also be contributed to a vessel risk analysis that, in combination with other data, can lead to vessels being given special attention when entering the littoral state and the port.

The remaining 90% of the AIS data, although significantly decimated, is relatively clean and can be used by the data analytics applications and when used by third party applications, provides more reliable information/reports.

5   Event Management

An event manager, consumes clean AIS data and processes the data to detect a wide range of events and then generate one of a number of alerts.

AIS data along with a range of geospatial data can then be used with greater confidence in an event manager to generate a range of standard events that then leads to the generation of additional data and the associated alerts. These include:

1. Port entry and exit events
2. Closest Point of Approach (CPA) violation
3. Events as vessels approach hazardous objects and areas

Within some environments, the events become far more complex and includes:

1. Threat determination when vessels sail through sensitive areas
2. Threat determination when vessels enter sensitive areas
3. Vessels in dangerous or environmentally sensitive areas
4. Detecting the interaction between ships and work boats with a port such as pilot boats, tug boats and bunkering vessels
5. Detecting unusual activity of high risk vessels such as high-speed ferries within a port environment
6. Detecting the interaction between vessels of interest at sea

Each of these types of events need to deal with a sequence of sub-events to enable the detection of the desired event with a high level of confidence that the event took place and then generate the alert that enables the operator to take the correct action timeously for that event in that area.

The large volume of data that is now available along with the processing power to be able to process this data and deal with the described events allows authorities the unique capability to determine the performance of any port (number of ships arriving, in port and departing each day) or any ship (average sailing speed, acceleration, deceleration and port calls on any on voyage) that was fitted with a Class A AIS unit.

Where the AIS and event manager data is stored over long periods of time often approaching several years, using analytics, seasonal and yearly trends in ship activity, port activity and shipping routes can be detected.

The detection of routes can be used a predictor of the route to be followed by a particular vessel to the next known destination port. Where any particular vessel or fleet of vessels has been tracked over a number of years, the prediction may use this historical data to add to the accuracy of the destination port and the route used by that vessel or vessel fleet to that particular port.

6 Pattern Recognition

A large number of ships often sail a predetermined and often repeated route and thus the routes are often simpler to detect.

The sailing patterns that are of great interest to many are the fishing fleets. This data is becoming more widely available because more fishing vessels are being fitted with AIS units and the size of the vessels being fitted with AIS is decreasing. Smaller vessels tend to fish closer to their home port and these fishers are often dependent on their catch for food as well as income. The number of small fisher vessel in some countries is significant and can approach more than 500,000 small vessels in some Asian countries.

The fishing community operates in a high-risk environment. In the short term, using pattern recognition techniques, the activity of ships and small craft fitted with AIS can be determined and measured against newly created models for safety, security, economic and environmental monitoring purposes.

The combination of catch data and voyage data allows the catch to be certified as being completed under the agreed terms and conditions applicable to that fisher or that fishing area and this, when attached to the catch can add economic value to the catch because of this catch source tracking capability.

7   Volume of Data

A typical national authority will monitor the activity of all 165,000 ships that can be detected by AIS in 2017. This monitoring is used to reduce the various risks when vessels enter the littoral state and national ports. These 165,000 vessels generated an estimated 19,000 AIS messages per second on a global basis resulting in 48 Mb/s data being transmitted in the two AIS channels. Many vessels are at sea and the transmitted AIS messages are not detected by terrestrial or satellite AIS receivers. About 8 million AIS messages are collected per day by space based AIS receiver systems with a revisit rates of about 15 min at some latitudes.

The number of vessels detected using AIS has increased by about 40% over the last 4 years due to the increased number of vessels fitted with AIS units, the improved performance of AIS receivers and antennas on the various satellite systems. This has increased the number of AIS messages received and the number of vessels.

In areas where the AIS population has increased significantly, techniques, such as coverage area sectorisation is being employed by terrestrial AIS receiver systems to gather more reliable data. A number of national authorities have now recognised that the monitoring of their entire coastline is important for safety, security, economic and environmental reasons and although not increasing the number of vessels being monitored, this activity is increasing the number of AIS messages for each vessel that are being collected, processed and stored on a daily basis.

AIS unit cost is decreasing and thus encouraging a large number of leisure and other small craft to fit AIS.

The following factors are expected to drive the amount of AIS data stored and available for analysis:

1. A decrease in AIS data decimation
2. The increasing number of satellites fitted with improved AIS receivers and antenna systems
3. The increasing number of work vessels, ferries, fishing and leisure craft being fitted with AIS
4. The integration of maritime locating systems and/or sensors to form an integrated system with a Common Operating Picture (COP) or Single Pane of Glass (SPG) operating environments

5. The growing AIS coverage of the various littoral states for safety, security, economic and environmental reasons
6. The inclusion of AIS as part the new and higher data bandwidth VHF Data Exchange System (VDES) technology
7. The sensorisation of ships and the cargo that is being carried by these ships (the Internet of Things (IoT))
8. The increase in applications that generate new data from the historical and real time AIS data available from various sources
9. Encouragement of the mechanised, artisanal and sport fishing environments to fit AIS for their own safety and protection

8 Satellite AIS

Satellite AIS was trialed as an experiment by ORBCOMM in the mid 2000's with a view to this being an opportunistic method of collecting AIS data on a global basis. This was proven to be a marginal but operational solution and since then, many satellites have been fitted with AIS receivers and placed in Low Earth Orbit (LEO) to collect AIS transmissions from ships, Aid to Navigation AIS and AIS shore stations.

AIS receivers that are used in space have become more sensitive and able to detect more AIS transmissions through deconflicting techniques. These orbiting LEO satellites have not provided simultaneous global coverage and use store and forward techniques leading to some system latency between when a ship transmitted the AIS message and the time this is made available to the consuming entity.

Various data processing and data analytic techniques have been used by the satellite service providers to extract and provide high quality data. This is then able to be used by their customers to enhance the national Maritime Domain Awareness (MDA) systems.

The pace of progress with regards to AIS receivers in space is not slowing down. Iridium are placing AIS receivers on more than 60 IridiumNext satellites and this system is expected to capture more than 50 million AIS messages per day with a revisit rate of less than 2 min on a global basis.

The near real-time availability of AIS data for all vessels fitted with AIS is driving new opportunities in port, berth and ship monitoring. The analysis of the ship voyage data is adding value to determining the risk that any vessel poses to a national authority or port and is also allowing those interested in the risk profile of any vessel or fleet of vessels to utilise the detailed spatiotemporal data that is now becoming available to re-evaluate and fine tune their risk models.

9 The AIS Maritime Information System Infrastructure

This volume of AIS and associated data is driving the centralisation of AIS and associated data and services into cloud based systems and services offering a global service.

There are a number of standards that apply to AIS network infrastructure of which the Common Shore Side Architecture (CSSA) is the most well-known.

The CSSA does not however, deal with performance required to process the various AIS data streams and then provide the various topical data streams known in AIS as Logical Shore Stations (LSSs) and then analyse the real-time (real time analytics), historical (historical analytics) and actionable AIS data (prescriptive analytics) along with the many events that have resulted from the real-time processing of the AIS data.

The first challenge is to collect the data from the various sources to satisfy the demands of the end user. This could include AIS data that originates from:

1. AIS receivers in the area of interest
2. AIS shore stations in the area of interest
3. AIS Aids to Navigation devices in the area of interest
4. Satellite AIS data on a national, regional or global basis
5. Data from third party sources such as:
   a. Neighbouring countries
   b. Port Vessel Traffic Services
   c. Off shore energy platforms
   d. IALAnet
   e. Maritime Safety and Security Information System (MSSIS)

This data is cleaned using streaming analytics to remove any AIS data that is detected as being corrupt and then merged into a single temporally sequenced database which serves two primary purposes:

1. Store the received and cleaned AIS data
2. Provide an audit trial of all AIS data from source to customer

The AIS data is then streamed according to the LSS filtering and processing requirements. During this process, the data is processed in parallel by the event manager and the target trajectory extractor.

The stored data is made available for historical analytics. All data is made available on line for reporting via a Human Machine Interface (HMI)/query builder or by using a web service.

The performance demanded by users is often given in response time for a particular query on historical data such as:

1. A 10 s response time is required from a query to data result being available for download for all AIS data and trajectory data for one single vessel for one month from any month in the last 10 years of stored AIS data. The download of the result data is excluded.
2. A 10 s response time is required from a query to data result being available for download for all vessels that crossed a dynamically defined 100NM by 100NM bounding box in the selected one month from any month in the last 10 years of stored AIS data. The download of the result data is excluded.

To achieve the required performances demanded by users requires a range of techniques that includes:

1. Storing of the latest near real-time data in volatile Random Access Memory (RAM) database (i.e. x minutes)
2. Use of Solid State Disk (SSD) storage for some often-used short term data (i.e. x days)
3. The sharding of the database using various well-known techniques
4. Running multiple copies of the same database
5. Using a high performance, low latency LAN
6. Predefining a range of reports that are run in the background
7. Preparing the often-used AIS data in the database in a manner that is optimised for most standard reports

Historical analytics with the large volumes of AIS data available is becoming of interest to the fishing industry when combined with weather predictions, meteorological and hydrological data with water temperature.

10   VDES: A Step Change in Maritime Data Services?

A VDES unit includes AIS, Application Specific Message (ASM) + VHF Data Exchange (VDE) RF technologies. The AIS technology data rate is 9.6 Kbps, the AMS technology data rate is 19,2 Kbps and the VDE technology data rate is 308 Kbps.

A VDE capable unit has 32 times the bandwidth of an AIS capable unit.

The VDES specification, ITU-R M.2092-0 was published in 2016 for terrestrial services after the Radio Frequencies were approved for use at WRC-15. At WRC-19, an application is being made for Radio Frequencies (RF) to enable VDES satellite communications on a global basis.

VDES allows ship to ship, ship to shore, shore to ship and ship to satellite communication. The interface and the ship and the shore side are based on the same standards as was used for AIS. VDES has several unique technical features that includes:

1. Dynamic bandwidth allocation
2. Forward Error Correction (FEC)
3. Selected message authentication adding a level of cyber security
4. Backward compatible with AIS

VDES is one of a number of technologies that is increasing the bandwidth available to ships that includes satellite communication systems in the deep sea and LTE communication systems when close to the coastline and ports in some parts of the world.

The VDES technology is being propelled by the communication demands of e-Navigation. VDES with associated data collection, data processing, data storage, event management and data analytic engines is expected to enable a wide range of new maritime e-Navigation centric services such as the Sea Traffic Management (STM) and Port Collaborative Decision Making (PCDM) systems that could be supported by a wide range of localised applications that could run on any number of portable communication devices within the maritime domain.

There are large numbers of sensors and applications that are expected to be accommodated on the new VDES system starting from early 2018 onwards.

11 Future Opportunities

Autonomous ships are now being considered and along with e-Navigation and the drive to sensorise ships and port environment. This results in significantly more location based and contextual data going to be available via the increasing data bandwidth to authorities and commercial companies that will drive a wide variety maritime centric safety, security, economic and environmental agendas. The availability of this increased volume of data along with an analytics capability will increase the view of the logistics chain from manufacturer in one part of the world to consumer in another.

With more finely grained AIS data becoming available due to the reduction in decimation and with AIS data being integrated with a wide range of IoT sensors in ports, on ships and fitted to cargo, weather data, radar data and data from other geospatial and imaging sensors, the volume and breadth of data to be collected, processed stored, analysed and reported on is growing and creating new insights into the maritime environment resulting in new commercial opportunities.

These new commercial opportunities include:

1. Streaming analytics to clean AIS data providing real time insight into vessel activities with complex events adding new data and new insights
2. Historical analytics allow the development of contextual models and patterns to provide insight and support the streaming analytics and allows the development of prescriptive analytics applications
3. Detection of complex events adding data and information available to support the maritime operator environment
4. The integration of IoT sensors on the ship and shore will support the improvement in efficiencies and lead to further gains in safety, security and improve the economic viability and the environmental protection in this domain

All changes, disruptions and enhancements to the operational maritime environment need to have a compelling moral, legal and/or commercial imperative or they will not take hold.

12 An Example That Demonstrates the Potential

We are going to work through an example that comes from the European Union H2020 EfficienSea2 project. This example follows a vessel from Port A to sea and from sea to Port B.

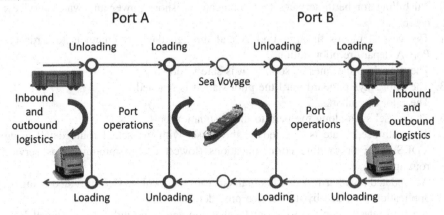

1. The vessel is berthed at Berth A1 in Port A its location is determined by two components:
   a. The size of the berth – this can be made available from the port database or determined from the historical data for vessels that have berthed in the same/similar position.
   b. The size of the vessel.
2. The vessel is being loaded with cargo as part of port operations.
3. The cargo will be in line with the ship and cargo type (i.e. Cargo vessel Carrying DG, HS, or MP, IMO hazard or pollutant category X).
4. The vessel has an estimated date and time of departure 3 h in the future. History indicates that this vessel leaves on time.
5. The master of the vessel obtains the best passage optimised for vessel type, predicted weather and sea state 4 days ahead.
6. The route is loaded onto the Electronic Chart System (ECS) and transmitted to the port control of the port, the vessel operator, the destination port and any Vessel Traffic System (VTS) in between the departure and destination port using the port data service (VDES, WiFi or LTE).
7. The port VTS is notified.
8. The tug boat operator is notified.
9. The pilot is notified.
10. The linesman is notified.
11. The loading is completed and port operations notified.
12. The ships agent is provided with a status update transferred to the shore using VDES or LTE.
13. The pilot boards and his costs start increasing.
14. The tug boat arrives and ties up and its costs start increasing.

15. The lineman arrives.
16. The port control gives permission for the vessel to leave.
17. The lineman releases the ship.
18. The tugs pull the ship from the berth.
19. The berth is now empty and can be used for another vessel.
20. All billing for berth services (communications, shore power and water services) cease.
21. The vessels moves through Port A and through the port channels towards the Port A departure pilot station.
22. The tugs drop the lines as soon as it is safe to do so.
23. The pilot stays onboard until the pilot station is reached.
24. The pilot disembarks.
25. The vessel sails along the path to the destination port.
26. The optimised route is updated on the vessel with the actual route detail using VDES, LTE or satellite communications (lowest cost communications service routing).
27. An updated weather forecast is obtained from the local service provided using an application provided by the service provider.
28. A new route is required to avoid local traffic and is provided by the local VTS office.
29. A new optimised route is constructed on the ship or by a service provider and sent to the vessel operator, the destination port and all VTS offices between the vessel and the destination port.
30. The destination port ETA data is entered into the destination Port Collaborative Decision Making (PCDM) system.
31. The vessel, if it maintains its current sailing speed, is going to arrive 3 h too early and an optimised route is sent back to the vessel to allow it to 'green steam' to the destination port saving fuel and reducing pollution.
32. Work flow systems provides the port operations with the list of actions to take to ensure all operational arrangements are completed.
33. The destination port VTS is notified.
34. The vessel arrives at the approach to the destination port, Port B.
35. The ships agent, pilot and tug boats are all notified and ready for the vessels arrival.
36. The pilot boards at the pilot stations.
37. The tugs tie up to the vessel at the designated point.
38. The tugs take the vessel to the designated berth at which the linesman is ready to secure the ship.
39. The ship enters the berth and the required services are made available (water, shore power and communications services).
40. The berth is marked as being occupied.
41. The linesman secures the ship.
42. The tugs depart.
43. The ships agent is provided with a status update of all tracked cargo using IoT and transferred to the shore using VDES or LTE.
44. The unloading operations begin.

45. The vessel has an estimated date and time of departure 3 h in the future. History indicates that this vessel leaves on time.
46. The master of the vessel obtains the best passage optimised for vessel type, predicted weather and sea state 4 days ahead.

..... and so on.

13 Summary

Data Analytics enables advanced AIS applications and the associated safety, security, economic and environmental protection applications through:

1. Ensuring that AIS data that is collected processed and stored is clean
2. Creating visual reports that can easily be viewed, interpreted and used by the operator
3. Using an event manager to generate new data from a wide range of preconfigured and operator configurable events
4. Allowing the operator to create reports that combine data from the AIS Comment Blocks, static, dynamic and communication state data
5. Take the load from the operator by implementing prescriptive analytics that uses all received and generated data
6. Including the Sea Traffic Management (STM) and the port operations Port Collaborative Decision Making (PCDM) applications
7. IMIS Global is a datAcron partner and is focused on the maritime domain using AIS data and the implementation of event management, data compression and analytics in the maritime domain.

References

1. datAcron project. http://www.datacron-project.eu/
2. ITU-R M.1371-5 specification. http://www.itu.int/rec/R-REC-M.1371-5-201402-I
3. AIS overview. http://www.iala-aism.org/product/an-overview-of-ais-1082/
4. VHF Data Exchange System overview. http://www.iala-aism.org/product/vhd-data-exchange-system-vdes-overview-1117/
5. VHF Data Exchange System conference proceedings, Cape Town, South Africa. http://www.vdesconference2017.co.za/presentations.html
6. EfficienSea2 project. http://efficiensea2.org/
7. Sea Traffic Management project. http://stmvalidation.eu/
8. Piracy reports and AIS data spoofing. http://piracyreport.co.za/The_Curious_Case_of_the_Hacker-Pirates.html
9. Port Collaboration Decision Making. http://stmvalidation.eu/activity-item/activity-1-port-collaborative-decision-making/
10. Orbcomm Inc. https://www.orbcomm.com/en/networks/satellite-ais
11. IridiumNext. https://www.iridium.com/network/iridiumnext

What do Geotagged Tweets Reveal About Mobility Behavior?

Pavlos Paraskevopoulos[1,2(✉)] and Themis Palpanas[2]

[1] George Mason University, Fairfax, USA
paraskevopoulospavlos@gmail.com
[2] LIPADE, Paris Descartes University, Paris, France
themis@mi.parisdescartes.fr

Abstract. People's attention tends to be drawn by important, or unique events, such as concerts, demonstrations, major football games, and others. Many individuals are even willing to travel long distances in order to attend events they regard as important. As a result, the everyday patterns that a person has, changes. This includes changes in the normal mobility patterns of this person, as well as changes in their social activities. In this work, we study these phenomena by analyzing the behavior of social media users. We investigate the activity and movement of users that either attend a unique event, or visit an important location, and contrast those to users that do not. Furthermore, based on the online activity of users that attend an event, we study the information that we can extract related to the mobility of these users. This information reveals some important characteristics that can be useful for a variety of location-based applications.

Keywords: Social networks · Social ties · Geolocation · Movement
Twitter

1 Introduction

The mobility of people and the reasons that cause them to move has been an interesting research topic [1–3] that could be used in order to lead to more efficient urban planning, and to a better understanding of human behavior with regards to unique events. Unique events, such as concerts or football games, tend to attract much attention, while many people are willing to travel long distances in order to attend them.

These events may also affect the social media activity of the people who attend them, as well as the activity observed in the areas they take place, an aspect that is used by some studies [4–7]. The observed increase in (geolocalized) posts on social media from the location a unique event takes place in [5], indicates that social media users tend to share with their friends moments that make them happy or excited, often times also sharing their locations, creating geotagged posts, contrary to their normal patterns. This increase of the social

© Springer International Publishing AG 2018
C. Doulkeridis et al. (Eds.): MATES 2017, LNCS 10731, pp. 36–53, 2018.
https://doi.org/10.1007/978-3-319-73521-4_3

media activity reveals the social ties [8] created between users, simply as a result of attending the same event.

In this work, we try to understand what forces users to make geotagged posts, by observing their mobility through the geotagged tweets. We also investigate if unique event attendants share normal activity and mobility patterns. Finally, we examine the number of the users needed to reveal some important character- istics such as routes or the shape of a country. In order to achieve our targets, we propose a set of methods, which we evaluate using a dataset consisting of geotagged posts from Twitter.

The contributions we make in this paper can be summarized as follows.

1. We employ user samples of different sizes, and study how the sample size affects the information on the most important mobility and activity patterns of users.
2. We examine the difference of the activity and mobility behavior of people who attend an important event, as opposed to the general user population, and show that attendance of certain events imply increased mobility for these users.
3. We present results indicating that user presence in special events or locations is related to the activity patterns of the user, and increases the likelihood of making geotagged posts.

The (preliminary) results of our analysis can be useful for a variety of appli- cations, such as in marketing. In this case, the advertisers can choose target groups also depending on their mobility characteristics, which can in turn be determined by knowing some specific locations and/or events that a user has visited.

2 Related Work

2.1 Social Media and Social Network Analysis

Smart devices give users the opportunity to use social media regardless of their location: house, office, or street. They also give the choice to the user to mark her position when posting a message or a photo, creating geotagged posts. In our study, we concentrate on the analysis of data that derive from Twitter. Twitter is a social network that gives to the user the chance to express feelings, or make comments, by using a 140-character text. Although only around 2% of all the tweets are geotagged [9,10], these are enough to extract important events and their locations, while also increase the volume of the geotagged information by geolocalizing non-geotagged posts [11].

A study that observes movement of people by checking geotagged tweets is presented by Balduini et al. [12]. In this study, the authors analyze geotagged tweets originated from London, and more precisely close to the Olympic sta- dium during the Olympic games, and identify the exact movement of the crowd, especially during the opening ceremony. Other works have also proposed tools for the analysis and visualization of such geotagged information [13–15].

Observing the movement of the crowd is very interesting, but it is not the only question that researchers have tried to address. Some studies focus on the extraction of local events by the analysis of the text posted in tweets. Such a study is presented by Abdelhaq et al. [16]. The target of this study is to identify local events. In order to achieve its target, it initially uses both geotagged and non-geotagged tweets for identifying keywords that best describe events. Then it keeps only the geotagged tweets and extracts the local events. Another interesting study that uses tweets in order to identify events and to explain social media activity during interesting events is presented in [17]. In [4] the authors try to identify where an earthquake happened by only analyzing the activity on Twitter.

Cho et al. [18] develop a framework for analyzing periodic and not periodic movement of the users of social networks, using data of social networks and mobile data. Another interesting study that identifies both aggregated mobility patterns and mobility patterns for unique users is presented in [19]. The authors of this paper use data from online social media such as Twitter and Facebook in order to get how the popularity of a location affects the destination of the user. In [20], Hu et al. present a method that targets to predict future location based on what a user posts, when it's posted and from which location. On the contrary to the previous methods that rely on Twitter or Facebook, Noulas et al. in their work presented at [21] analyze the spatio-temporal patterns of the users' activity and their dynamics using check-ins from Foursquare.

In [22], Crandall et al. investigate the social-ties two users have, based on the co-occurrences they have at a set of different locations. In order to achieve this, they apply a spatio-temporal probabilistic analysis on geotagged photos collected by Flickr. Finally, a study that analyzes the demographics of the people who participate at the movement "#blacklivesmatter" is presented by Olteanu et al. [23]. In this study, the authors investigate the demographics of the users, creating groups of the users based on their activity. On the contrary to this study, we analyze the activity of the users taking into consideration only their id and their geotagged tweets, achieving a more privacy aware analysis.

We note that the studies mentioned above either do not analyze movement, or if they do so, it is at the granularity of a city. Our target is to analyze the mobility caused by unique events at a country level. Furthermore, we study the characteristics of the activity and mobility patterns of different users, and how these are affected by unique events.

2.2   Studies on the Mobility of the Users

Apart from the studies based on social media and social networks, there are also several studies related to mobile phone usage data, GPS devices, or even bank note distribution, aiming to predict the mobility of users, or to analyze the differences of the activity of an area, based on user movement.

Ashbrook et al. [1] present a two level model that applies a clustering at the location recorded by car GPS devices and a probabilistic model in order to find the next location of the user. The target of this study is to identify the most important locations, while also predicting the movement of users. GPS traces are also used by Krumm et al. [2]. In this work, the authors present their algorithm, which uses a probabilistic model and historical data in order to predict the destination of the user, while also identifying deviations from the user's normal patterns. Do et al. [24] present a probabilistic model that predicts the location of a user at a future time interval, by using GPS data from smartphone devices. The study presented in [25] identifies mobility patterns based on trajectories that are created from anonymized mobile phone users and the travel distances of each user. The authors of [3] on the other hand, identify a set of features using a supervised method on GPS data, extracting mobility patterns.

Most of the studies previously described operate on GPS data. The study presented by Scellato et al. [26] proposes the "NextPlace" framework, which operates using either GPS or WiFi data. The target is to identify the location of a user, based on a spatio-temporal analysis of the data of the network. Chatzimilioudis et al. [27] present a set of algorithms that use trajectories for achieving a crowdsourcing analysis. Their framework can be applied in both outdoor and indoor environments, while their results target to help in cases such as minimizing of energy consumption of networks. Thanks to the mobile devices, users can call or send messages any time, creating Call Detail Records (CDRs). A study that uses CDRs is presented in [7], targeting to identify mobility patterns, while also explain the differences of the activity of a location based on CDRs and an event dataset. This study operates on cell-tower granularity. Other studies use data that are not so obvious they can reveal the users' mobility. Such an example is the study presented by Brockmann et al. [28] that targets to identify users' mobility by applying a spatio-temporal analysis of the banknote distribution.

All the methods described above target to predict user mobility based on either the individual user's patterns, or on some identified general patterns. In our study, we focus on the analysis of the mobility *differences* between groups of users that share some special characteristics (such as a common location at a specific time interval). As a result, we study deviations of normal patterns, and the reasons these deviations appear. Furthermore, the methods previously described operate on datasets where each user has a lot of points. In our study, in order to achieve our movement analysis, we use geotagged posts from Twitter, which results in a very sparse dataset (a user can have just 1–2 geotagged tweets in a period of 4 months), limiting the amount of available information.

3   Proposed Approach

3.1   Problem Description

The problem we want to investigate in this study is the identification of differences in the social media activity between users that attended an important event (e.g., a concert) and those who did not. In addition, we want to study the reasons that force a user to generate a geotagged message, as well as the corresponding mobility patterns. Finally, we would like to examine the extraction of

---

Algorithm 1. Get Representative Sample and Characteristics

INPUT: Temporal and Spatial parameters.
OUTPUT: A representative sample of users and its activity and movement.
1: $P_{WinInterest}, Q_{WinInterest} \leftarrow GetUsers(FGL, win_{ev}, CGL, WinInterest)$ ▷ get the users from the event *location* and the *CGL* and their activity
2: $users, activity, movement \leftarrow$ Percentage of top uses in P, Q ▷ get the representative users' sample
3: **return** *users, activity, movement*

---

a sample of users in a social network, that could allow us to reproduce the main routes that the users follow.

In the context of this work, we concentrate on users who attend major events or sights, such as concerts, or an important touristic attraction. Furthermore, we focus on Twitter, a social network that has more than 313M users, 80% of which are on mobile devices[1].

3.2 Methodology

In this section, we describe the method we developed for tackling the problems previously described (for the general schema, refer to Algorithm 1). Our method is based on the creation of social ties [8], where as social ties we define the connection between users that may not have common characteristics, except for visiting (independently) the same location during a specific time period. Initially we set the temporal and spatial parameters we are interested in. We then remove the spam and bot accounts based on the observed activity of the account. Finally, we follow the geolocalized posts users send during a predefined period of time. In the following sections, we elaborate on the methods discussed above.

Setting the temporal and spatial parameters. We start by setting the temporal and spatial parameters we are interested in:

1. FGL: the location the event is going to take place in (Fine-Grain location)
2. win_{ev}: the time window during which users visited FGL
3. CGL: the Coarse-Grain location (e.g., city, country) in which we will observe the movement of users
4. $WinInterest$: the period of time we will follow the users' geotagged posts

Get the Event and CGL users.

In order to get the initial sample of our users, we use the spatio-temporal parameters and we check our dataset for users that posted at least one geotagged tweet from the event location, before, during or after the event (win_{ev}). Afterwards, we get all the geotagged tweets these users posted, for a predefined period of time ($WinInterest$).

Having already extracted the users who attended the event, we get the rest of the users from our CGL that have at least one geotagged tweet during the

---

[1] https://about.twitter.com/company.

Algorithm 2. Get Users

1: **procedure** GETUSERS($FGL, win_{ev}, CGL, WinInterest$)
2:     $U_{FGL,win_{ev}} \leftarrow$ all users at FGL at time-window win_{ev}
3:     **for all** $u \in \{U_{FGL,win_{ev}}\}$ **do**   ▷ get first sample of users in FGL and their activity
4:         $P^{u,CGL}_{WinInterest} \leftarrow$ all tweets user u posted from CGL during time-window $WinInterest$
5:     $U_{CGL,win_{ev}} \leftarrow$ all users at CGL at time-window win_{ev}
6:     **for all** $u \in \{U_{CGL,win_{ev}}\}$ **do**   ▷ get all users in CGL and their activity
7:         **if** u not in $U_{FGL,win_{ev}}$ **then**
8:             $Q^{u,CGL}_{WinInterest} \leftarrow$ all tweets from user u at time-window $WinInterest$
9:     $P_{WinInterest} \leftarrow$ SpamFilter($P^{CGL}_{WinInterest}$) ▷ clean spam and bot accounts from $P^{CGL}_{WinInterest}$
10:     $Q_{WinInterest} \leftarrow$ SpamFilter($Q^{CGL}_{WinInterest}$)   ▷ clean spam and bot accounts from $Q^{CGL}_{WinInterest}$
11: **return** $P_{WinInterest}, Q_{WinInterest}$

---

win_{ev} and they have no tweets from the FGL during this time interval. The steps that we follow in order to get the users we are interested in, are presented in Algorithm 2.

Data Cleaning.

There are a lot of accounts that are either bots sending posts with the same content for a long period of time, or accounts that are sending posts with different content, from the exact same location. These accounts do not offer any useful information for our problem (they actually induce noise), therefore, we filter them out. More specifically, for a given account we check if at least 30% of the posted messages have the same prefix, latitude, or longitude. If an account meets at least two of the three conditions, we filter out the account.

Activity and Movement Comparison.

After the extraction of the datasets of the location place and the CGL, we compare their activity using the cumulative distribution function (CDF). Using the CDF, we can compare the activity between the users who visited the event locations and those who did not. Furthermore, we check the distribution of the other locations they visited during the time interval of interest, $WinInterest$. In order to achieve this, we compare the difference between the maximum and minimum latitude and longitude the user appeared in.

The hypothesis we want to verify using the above analysis is that users tend to travel long distances in order to visit a unique event or a unique location. In addition, we want to verify a second hypothesis, that users are more willing to share their location in case they attend important events, as opposed to their normal activity patterns.

4   Experimental Analysis

In order to evaluate our ideas, we used geotagged posts from Twitter. The datasets we used contain events such as major concerts and important touristic locations. In this section, we present a set of activity and movement analytics, while we provide the reader with visualizations of the location we get the tweets from.

Dataset Description

For the evaluation of our methods, we used a dataset containing geotagged tweets generated from Italy (as defined by a bounding box) for the period between 1st of June and 20th of October 2016. In this dataset, we have 1,460,083 geotagged tweets, posted by 173,182 unique users. We focused on important locations and events that took place during these time intervals. More precisely, we targeted users who posted geotagged posts from *Vatican* in Rome, and the concert of Bruce Springsteen in Rome, which in our experiments is referred to as *Concert*.

4.1   Important Event and Location Activity Analysis

People Attending *Concert*

We initially focused on an important event that took place in Rome and attracted a lot of people. This event was the *Concert* that took place at the location "Circus Maximus" on 16 of July 2016. We found the users that visited this location and posted a geotagged post since the midnight of the previous day. The time windows that we used were 24 h and 48 h (it was a 2-day concert), searching for posts initially posted up to the end of the concert (i.e. 24 h) and afterwards also the following day (i.e. 48 h). Having identified the users who generated messages from this location during our window, we followed all their geotagged posts for the period between June 1st and October 20th, 2016.

After further analyzing the activity of these users, we found that it was a sample of 67 non-spamming users[2]. The statistics of these users' activity is presented in Table 1. As we can see in this table, when decreasing the number of users in our sample, keeping a percentage of the most active ones, the standard deviation of the activity of the users is not affected much, while the mean activity of the users decreases. This fact implies that the distribution of the activity of the users is similar for the majority of the users in our sample, and especially for the most active users.

In Fig. 1a we depict the locations these 67 users "appeared" at, while in Figs. 1b, c, d we can see respectively the locations the 75%, 50% and 25% most active users posted geotagged tweets from, for the period June to October. In all the plots we present in this section, each color represents a different user[3]. As we can see in Fig. 1a, the combination of mobility and activity patterns of these 67 users cover the entire country of Italy. Furthermore, after manually checking the position of highways in Italy, we found out that these 67 users are able to form

---

[2] All users reported in the experimental part, are non-spamming users.

[3] Due to the relatively high number of users, different users may share the same color.

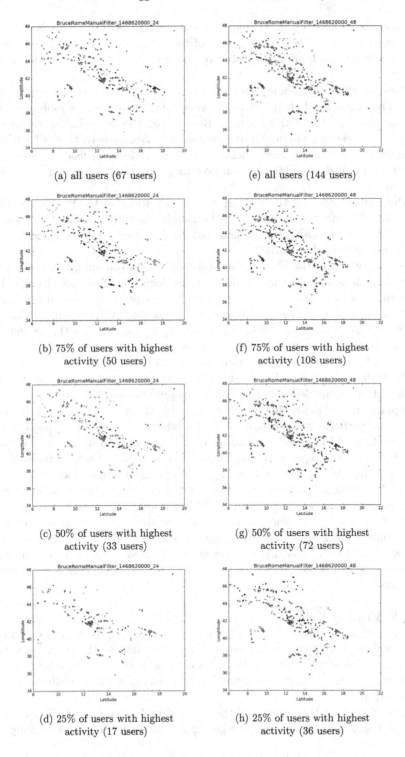

(a) all users (67 users)

(e) all users (144 users)

(b) 75% of users with highest activity (50 users)

(f) 75% of users with highest activity (108 users)

(c) 50% of users with highest activity (33 users)

(g) 50% of users with highest activity (72 users)

(d) 25% of users with highest activity (17 users)

(h) 25% of users with highest activity (36 users)

Fig. 1. *Concert*, 24-h (left) and 48-h (right) windows

the main shape and the main routes of the country of Italy. This is still true when we consider the 50% most active of these users (see Fig. 1c), and almost true even when we limit the number of the users to 17 (25% of most active, Fig. 1d).

These results reveal some very interesting characteristics of our dataset and users. They indicate that an extremely small number of users is mobile enough in order to cover the entire country. Recall that the users in the sample we examined belong to a particular demographics group, namely, they all attended a specific music concert. Nevertheless, this observation can lead to interesting marketing applications, since we can now target users with particular mobility patterns.

After having checked the activity and the locations of the people identified using the 24-h window, we analyzed the people identified by the 48-h window. The volume of the sample was increased to 144 users. The statistics of this sample are presented in Table 2.

As we noticed in the case of the 24-h window, sub-sampling with the most active users does not affect much the standard deviation of the activity. Furthermore, the mean activity is slightly decreased compared to the one of the case of the 24-h window, while the standard deviation is similar. This implies that the activity of the 68 users identified at the concert location during the second day, does not differ to the activity of the users of the first day.

In Fig. 1e, we can see the locations of the 144 users identified at the concert for the 48-h window. The fact that we increased the window, appending users to our dataset, provided us with more geotagged tweets. Due to this, we have more points in our plots, showing more precisely the map of Italy and the main highways. Furthermore, comparing Figs. 1c (which is formed by 33 users) and h (which is formed by 36 users) we notice that the shape of Italy formed by the 36 users is much more representative. This is due to the fact that the users, whose activity is depicted in Fig. 1h, have in general (slightly) higher activity.

Finally, in order to check the impact of the concert to the area, we slightly modified our parameters, targeting users that visited the concert area one week before the concert took place. Even though the area is located in the center of Rome, only 6 users had posted geotagged messages from this location during a 24-h window. This means that the concert was indeed the reason that the users posted geotagged posts (as we also verified by further analyzing the content of the posts).

People Visiting *Vatican*

Having analyzed the activity of the users who attended an important unique event such as a concert, we turned our focus on one of the most important locations of Rome, the *Vatican*. We followed exactly the same procedure we did in the case of the concert, modifying only the location whose visitors we were interested in.

After analyzing the activity of the visitors' of *Vatican* using the 24-h window, we found that 48 users posted geotagged tweets from *Vatican* during this

Table 1. Statistics for most active users of *Concert* and *Vatican* for 1 day

User%	Concert (1 day)				Vatican (1 day)			
	Users	Act%	Mean	Std	Users	Act%	Mean	Std
100%	67	100%	25	31	48	100%	23	42
75%	50	98%	33	32	36	98%	30	47
50%	33	91%	45	32	24	91%	42	54
25%	17	69%	68	33	12	75%	69	69

Table 2. Statistics for most active users of *Concert* and *Vatican* for 2 days

User%	Concert (2 days)				Vatican (2 days)			
	Users	Act%	Mean	Std	Users	Act%	Mean	Std
100%	144	100%	19	27	91	100%	25	56
75%	108	98%	25	29	68	95%	32	64
50%	72	95%	36	31	46	91%	45	74
25%	36	75%	57	33	23	79%	79	96

window. The activity of these users is presented in Table 1. Contrary to the case of the *Concert*, the standard deviation of the activity of the users that visited *Vatican* is affected when limiting the sample to the most active users.

In Fig. 2, we depict the locations of the users that visited *Vatican*, the same day that the concert was, and posted a geotagged post from *Vatican* for a 24-h and 48-h window, consecutively, while in Table 2 we present the statistics of the user sample we took using the 48-h window. As opposed to the case of the users who attended the *Concert*, the shape of the map of Italy that is formed is not very clear. This difference is more obvious when comparing the Fig. 1b, which was created using a sample of 50 users, with the Fig. 2a, which is created using a sample of 48 users. In the case of the 48-h window, this comparison is possible between the Figs. 1f (108 users) and e (91 users). Possible explanations for this behavior include the fact that the majority of the users who visited *Vatican* are tourists whose home-location is outside of Italy. Nevertheless, these results highlight the different mobility and activity behaviors of these two different samples of users.

4.2 Most Active and Random Users from Italy and Rome

Following the analysis of the users who either attended an event (i.e., concert) or visited an important location, we wanted to compare their activity with the people who haven't been located in one of the previous cases. In order to achieve our target, we have identified and followed users either from Italy or from (only) Rome that at the day of the *Concert*, were not located at the location the *Concert* took place. In order to make the comparison fair, we keep only n

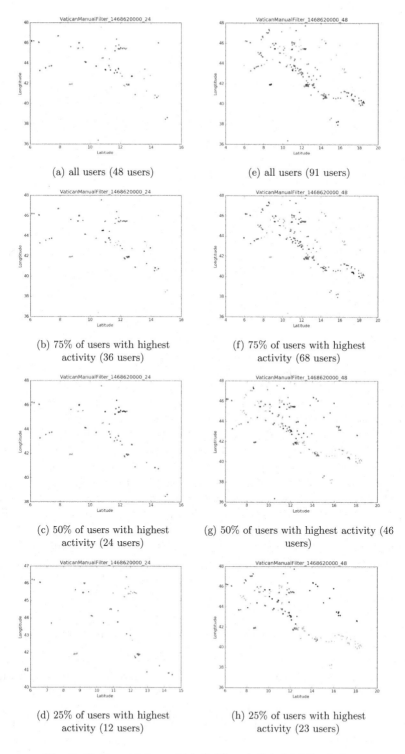

(a) all users (48 users)

(e) all users (91 users)

(b) 75% of users with highest activity (36 users)

(f) 75% of users with highest activity (68 users)

(c) 50% of users with highest activity (24 users)

(g) 50% of users with highest activity (46 users)

(d) 25% of users with highest activity (12 users)

(h) 25% of users with highest activity (23 users)

Fig. 2. *Vatican* visitors, 24-h (left) and 48-h (right) windows

users, where n is the volume of the sample of users who attended the *Concert*. Finally, having extracted the appropriate number of the users to be used, we experimented with the cases of the n Random people from the *Concert* and Italy/Rome, and Top n users from the *Concert* and Italy/Rome.

We note that after having manually analyzed the user activity, we found out that there is a single (non-spam) user with 26,756 distinct tweets (posts and replies to other users with his location identification "on"). We did not remove this user from the analysis.

Rome Visitors

In this part, we present the plots with the comparison of the locations between the Random n and Top n users from *Concert* and Rome.

We initially use 50% of the users that posted geotagged tweets from the location of the *Concert*, resulting in 33 users. The locations for the Top 33 and Random 33 users are depicted in Figs. 3a and b, respectively. As we can see in the figure, the representation of both the routes and the map of Italy is very accurate compared to the real map of Italy, despite the small number of users our sample has. As expected, when we increase the number of users to 67 (100% of the users that attended the concert), the representation of the map and the routes become even more clear (refer to Fig. 4). This is due to the fact that we have more users, and as a result, more tweets.

When examining the locations from which users posted geotagged tweets, we observe that the locations of the users that attended the *Concert* are more spread out than the locations of the users that posted geotagged tweets from Rome. This supports the assumption we made earlier, that users travel from more distant locations in order to attend a unique event such as a concert.

Italy Visitors

Having compared the activity and the location between the users of Rome and those of *Concert*, we wanted to compare the Top n and Random n users from Italy and *Concert*. Similarly to the case of the comparison of the two groups of users for Rome and *Concert*, the representation of the map of Italy and the routes are depicted clearly in the case of the 33 most active users of Italy (see Fig. 3e). Contrary to the case of the users from Rome, the result we get when using the 33 random users from Italy, depicted in Fig. 3f, is more representative of the country outline.

As depicted in Fig. 4, the representation of the routes becomes more clear when increasing the number of the most active users of Italy to 67.

4.3   Cumulative Distribution Function and Movement

In this part, we investigate the cumulative distribution function and the movement of the users who attended *Concert*, as opposed to those who did not.

As we can see in Fig. 5, the activity of the users who attended the concert differs from the activity of the users of Italy. More precisely, the percentage of the users who attended the concert and has a unique tweet is double compared

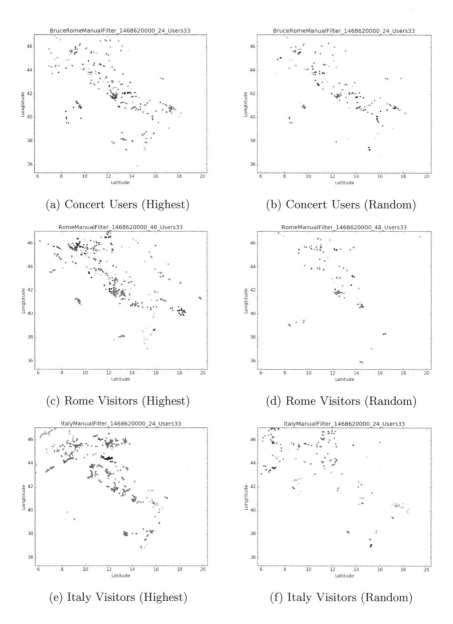

(a) Concert Users (Highest)   (b) Concert Users (Random)

(c) Rome Visitors (Highest)   (d) Rome Visitors (Random)

(e) Italy Visitors (Highest)   (f) Italy Visitors (Random)

Fig. 3. *Concert*, Rome and Italy visitors (top 33 and random 33 users)

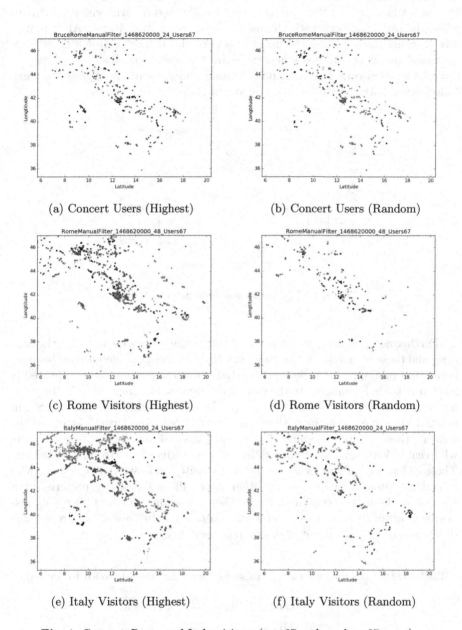

(a) Concert Users (Highest)

(b) Concert Users (Random)

(c) Rome Visitors (Highest)

(d) Rome Visitors (Random)

(e) Italy Visitors (Highest)

(f) Italy Visitors (Random)

Fig. 4. *Concert*, Rome and Italy visitors (top 67 and random 67 users)

to the percentage of the users who were located in Italy, but not in the concert area. Furthermore, we notice that the cumulative distribution function (CDF) of the users who attended the concert is very similar to those who visited *Vatican*, while the same happens with the users who were visiting Rome. After manually checking the tweets of the users who were at *Vatican* or Rome in general, we found out that the posts generated by the users who had posted only a few geotagged tweets, had been posted from unique locations such as *Vatican*, Colosseum or other historical monuments of Rome.

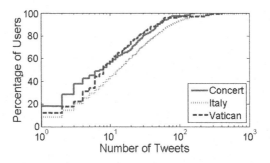

Fig. 5. CDF: Comparison of number of tweets

Furthermore, we compared the movement of the users who attended the concerts and those who did not. We found out that the median difference of the maximum and minimum latitude and longitude that the concert users appeared is 247 km in both dimensions. In the case of the users who were located in Italy but not at the concert the median of the latitude and longitude difference is 181 km and 231 km respectively. Regarding the users of Rome who haven't attended the concert, these numbers become 209 km and 228 km. Finally, regarding the users who visited *Vatican*, the median differences get reduced to 218 km and 163 km. These differences of the locations the users visited indicate that the users who intend to attend a unique event, either have different mobility patterns than the rest of the users, traveling a lot, or they are willing to travel from long distances in order to attend an event such as a concert. In Table 3, we present the differences of the movement each group of users had.

Table 3. Distances between the furthest locations that users traveled to (in km)

	Lat Dif Median	Lon Dif Median
Concert	247	247
Vatican	218	163
Rome	209	228
Italy	181	231

The difference at the locations the users of each group appeared, constitutes one more hint, reinforcing our initial hypothesis that users who attend important unique events, such as concerts, tend to travel from other locations in order to attend the event. Furthermore, these numbers combined with the distribution of the locations the attendees of a concert appear, indicates that the users who attend unique events also tend to travel more.

4.4   Discussion

Overall, our results show that the movement of the users who attend a unique event (i.e. *Concert*) is much higher than the rest of the users, indicating that the attendants of a unique event are willing to travel from long distances in order to attend the event. The differences at the mobility between those who attend a unique event and those that do not is so high that by depicting the locations where 33 attendees tweeted from during a period of 4 months, is enough to reveal the shape of a country and its highways.

Furthermore, the analysis of the activity patterns of the users, indicate that even though the sample of the users who attend an event shares only one common characteristic (i.e. attended an event), their activities follow specific patterns. The cumulative distribution function indicates that a sizable percentage of the users (around 20%) is willing to post geotagged information from the location the event takes place, which is opposite to their normal patterns. The analysis of the activity of the concert locations before and during the concert, reveals the effect that a unique event has to the activity of that location.

The differences between the activity and mobility patterns of the users who visited, or attended distinct places and events could be used in several applications, such as conducting marketing campaigns. For example, we could use the above information as one of the training features for designing targeted advertisements: focus on *Concert* attendees that travel long distances in order to advertise with a country-wide coverage (e.g., electronic devices), or focus on *Vatican* visitors that are mostly local visitors or tourists from abroad, who are going to explore the city, in order to advertise offerings that refer to the city (e.g., a restaurant).

5   Conclusions

In this work, we presented an analysis of the differences of the activity and mobility patterns of people that attend a major event, or visit an important location. Our results indicate that users are willing to travel from far locations in order to attend a unique event. Furthermore, we investigated the number of users needed to identify main routes and locations that attract people. This led to the surprising observation that the mobility and Twitter activity of less than 35 users that attended a unique event is enough in order to shape the main routes and outlines of regions, or countries. Finally, our experimental analysis shows that user presence in special events, or locations (such as an important touristic attraction, or a major concert) affects the normal activity patterns, increasing the likelihood of making geotagged posts.

As future research directions, it would be interesting to extend this analysis to a larger and more diverse set of locations, events, and time periods, verifying the results with different countries and event types. Furthermore, we would like to investigate how these results are affected by other factors, such as demographics (e.g., the age of the people who attend an event), time of year (e.g., during holidays), and accessibility of a particular location.

References

1. Ashbrook, D., Starner, T.: Using GPS to learn significant locations and predict movement across multiple users. Pers. Ubiquit. Comput. **7**(5), 275–286 (2003)
2. Krumm, J., Horvitz, E.: Predestination: Inferring destinations from partial trajectories. In: Dourish, P., Friday, A. (eds.) UbiComp 2006. LNCS, vol. 4206, pp. 243–260. Springer, Heidelberg (2006). https://doi.org/10.1007/11853565_15
3. Zheng, Y., Li, Q., Chen, Y., Xie, X., Ma, W.-Y.: Understanding mobility based on GPS data. In: Proceedings of the 10th International Conference on Ubiquitous Computing, pp. 312–321. ACM (2008)
4. Sakaki, T., Okazaki, M., Matsuo, Y.: Earthquake shakes twitter users: real-time event detection by social sensors. In: Proceedings of the 19th International Conference on World Wide Web, pp. 851–860. ACM (2010)
5. Li, R., Lei, K.H., Khadiwala, R., Chang, K.C.-C.: TEDAS: A Twitter-based event detection and analysis system. In: 2012 IEEE 28th International Conference on Data Engineering (ICDE), pp. 1273–1276. IEEE (2012)
6. Paraskevopoulos, P., Palpanas, T.: Where has this tweet come from? fast and fine-grained geolocalization of non-geotagged tweets. Soc. Netw. Anal. Min. **6**(1), 89 (2016)
7. Paraskevopoulos, P., Dinh, T.-C., Dashdorj, Z., Palpanas, T., Serafini, L.: Identification and characterization of human behavior patterns from mobile phone data. In: Proceedings of NetMob (2013)
8. Wuchty, S.: What is a social tie? Proc. Natl. Acad. Sci. **106**(36), 15099–15100 (2009)
9. Leetaru, K., Wang, S., Cao, G., Padmanabhan, A., Shook, E.: Mapping the global twitter heartbeat: The geography of Twitter. First Monday 18(5) (2013)
10. Murdock, V.: Your mileage may vary: on the limits of social media. SIGSPATIAL Spec. **3**(2), 62–66 (2011)
11. Paraskevopoulos, P., Palpanas, T.: Fine-grained geolocalisation of non-geotagged tweets. In: Proceedings of the 2015 IEEE/ACM International Conference on Advances in Social Networks Analysis and Mining 2015, pp. 105–112. ACM (2015)
12. Balduini, M., Della Valle, E., Dell'Aglio, D., Tsytsarau, M., Palpanas, T., Confalonieri, C.: Social listening of city scale events using the streaming linked data framework. In: Alani, H., Kagal, L., Fokoue, A., Groth, P., Biemann, C., Parreira, J.X., Aroyo, L., Noy, N., Welty, C., Janowicz, K. (eds.) ISWC 2013. LNCS, vol. 8219, pp. 1–16. Springer, Heidelberg (2013). https://doi.org/10.1007/978-3-642-41338-4_1
13. Paraskevopoulos, P., Pellegrini, G., Palpanas, T.: TweeLoC: A system for geolocalizing tweets at fine-grain. In: IEEE International Conference on Data Mining (ICDM) (2017)

14. City pulse. http://www.ict-citypulse.eu
15. Paraskevopoulos, P., Pellegrini, G., Palpanas, T.: When a tweet finds its place: Fine-grained tweet geolocalisation. In: International Workshop on Data Science for Social Good (SoGood), in Conjunction with the European Conference on Machine Learning and Principles and Practice of Knowledge Discovery (ECML PKDD) (2016)
16. Abdelhaq, H., Sengstock, C., Gertz, M.: Eventweet: Online localized event detection from twitter. Proc. VLDB Endow. **6**(12), 1326–1329 (2013)
17. Earle, P.S., Bowden, D.C., Guy, M.: Twitter earthquake detection earthquake monitoring in a social world. Ann. Geophys. **54**(6), 1–8 (2012)
18. Cho, E., Myers, S.A., Leskovec, J.: Friendship and mobility: user movement in location-based social networks. In: Proceedings of the 17th ACM SIGKDD International Conference on Knowledge Discovery and Data Mining, pp. 1082–1090. ACM (2011)
19. Hasan, S., Zhan, X., Ukkusuri, S.V.: Understanding urban human activity and mobility patterns using large-scale location-based data from online social media. In: Proceedings of the 2nd ACM SIGKDD International Workshop on Urban Computing, p. 6. ACM (2013)
20. Hu, B., Ester, M.: Spatial topic modeling in online social media for location recommendation. In: Proceedings of the 7th ACM Conference on Recommender Systems, pp. 25–32. ACM (2013)
21. Noulas, A., Scellato, S., Mascolo, C., Pontil, M.: An empirical study of geographic user activity patterns in foursquare. ICWSM **11**, 70–573 (2011)
22. Crandall, D.J., Backstrom, L., Cosley, D., Suri, S., Huttenlocher, D., Kleinberg, J.: Inferring social ties from geographic coincidences. Proc. Natl. Acad. Sci. **107**(52), 22436–22441 (2010)
23. Olteanu, A., Weber, I., Gatica-Perez, D.: Characterizing the demographics behind the #blacklivesmatter movement. In: 2016 AAAI Spring Symposium Series (2016)
24. Do, T.M.T., Dousse, O., Miettinen, M., Gatica-Perez, D.: A probabilistic kernel method for human mobility prediction with smartphones. Pervasive Mobile Comput. **20**, 13–28 (2015)
25. Gonzalez, M.C., Hidalgo, C.A., Barabasi, A.-L.: Understanding individual human mobility patterns. Nature **453**(7196), 779–782 (2008)
26. Scellato, S., Musolesi, M., Mascolo, C., Latora, V., Campbell, A.T.: NextPlace: A spatio-temporal prediction framework for pervasive systems. In: Lyons, K., Hightower, J., Huang, E.M. (eds.) Pervasive 2011. LNCS, vol. 6696, pp. 152–169. Springer, Heidelberg (2011). https://doi.org/10.1007/978-3-642-21726-5_10
27. Chatzimilioudis, G., Konstantinidis, A., Laoudias, C., Zeinalipour-Yazti, D.: Crowdsourcing with smartphones. IEEE Internet Comput. **16**(5), 36–44 (2012)
28. Brockmann, D., Hufnagel, L., Geisel, T.: The scaling laws of human travel. Nature **439**(7075), 462–465 (2006)

Edge Representation Learning for Community Detection in Large Scale Information Networks

Suxue Li[1], Haixia Zhang[1], Dalei Wu[2], Chuanting Zhang[1],
and Dongfeng Yuan[1(✉)]

[1] Shandong Provincial Key Laboratory of Wireless Communication Technologies,
Shandong University, Jinan 250100, Shandong, China
li_suxue@163.com, {haixia.zhang,dfyuan}@sdu.edu.cn,
chuanting.zhang@gmail.com
[2] The University of Tennessee at Chattanooga, Chattanooga, TN 37403, USA
dalei-wu@utc.edu

Abstract. It is found that networks in real world divide naturally into communities or modules. Many community detection algorithms have been developed to uncover the community structure in networks. However, most of them focus on non-overlapping communities and the applicability of these work is limited when it comes to real world networks, which inherently are overlapping in most cases, e.g. Facebook and Weibo. In this paper, we propose an overlapping community detection algorithm based on edge representation learning. Firstly, we sample a series of edge sequences using random walks on graph, then a mapping function from edge to feature vectors is automatically learned in an unsupervised way. At last we employ the traditional clustering algorithms, e.g. K-means and its variants, on the learned representations to carry out community detection. To demonstrate the effectiveness of our proposed method, extensive experiments are conducted on a group of synthetic networks and two real world networks with ground truth. Experiment results show that our proposed method outperforms traditional algorithms in terms of evaluation metrics.

Keywords: Network · Community detection
Representation learning · Cluster

1 Introduction

Networks that represent real world systems are everywhere in human life, such as biology, sociology, computer science, and academics [5,15]. An information network represents an abstraction of the real world, it provides us a useful tool to mine knowledge from links in it. Network analysis helps people solve real life problems. The study of networks is of great importance and attracts a lot of experts into it. One of the most important properties of network is community structure. The main purpose of community detection is to uncover the inherent structure of networks. Networks in real world are always complicated

C. Doulkeridis et al. (Eds.): MATES 2017, LNCS 10731, pp. 54–72, 2018.
https://doi.org/10.1007/978-3-319-73521-4_4

and large-scaled. It is hard to process and analyze it directly. By community detection techniques, networks are divided into different parts, the structure becomes obvious and the networks are more understandable which is helpful for subsequent analysis. Finding communities becomes a critical issue to study and explore networks. There are some typical applications listed as follows:

- **Advertising.** People in the same community often share similar interests. If we know a person buys a product, we can post advertisements about similar products to members who are in the same community with her. In this way, it can help improve the performance of product recommendation system.
- **Recommendation.** In social networking software, e.g. Facebook and Weibo, based on the community structure, we can recommend to a user those who are in the same community but not her friends yet.
- **Information propagation and Disease diffusion.** For overlapping community detection, we can find the fuzzy nodes that belong to more than one communities. Finding this kind of nodes is critical to speed up information propagation and control disease diffusion.
- **Information retrieval.** Words with similar meaning tend to be in the same community. When a user searches a keyword, the results of the keyword and its near-synonym can be presented to the user simultaneously. Thus community detection helps promote personalized services.

The majority of existing community detection algorithms focus on finding non-overlapping communities. However networks in real world are complicated and nodes in them often belong to many different communities, especially in social networks. In social networks, this kind of overlapping communities reflect different types of associations between people. [6] For example, two users in Weibo could be relatives, classmates, sport partners, etc. User A could be a classmate of B and a sport partner of C at the same time, so she should be in the community of classmates of B and in the community of sport partners of C simultaneously. What's more, A is important in the network, because B's classmates and C' sport partners can build connection through A. As shown in Fig. 1, there are two communities which both have four members. It is obvious that there is one fuzzy node belongs to both community 1 and community 2. This node plays an important role in this network, it serves as a bridge between two communities. When messages are propagated among different communities in a network, the propagation path will always go through this kind of fuzzy nodes. Thus finding such nodes which is called overlapping community detection is a critical issue in community detection. For this reason, it is an effective way to recover overlapping communities by grouping edges rather than vertices. Each edge is assigned to a single community, but clusters can still overlap because edges in different communities may share one endpoint.

In recent years, some algorithms based on machine learning methods appear. The architecture of these methods is shown in Fig. 2. Given a network, the first thing is to extract the node feature representations. Then the similarity calculation part is done. There are a variety of similarity calculation methods,

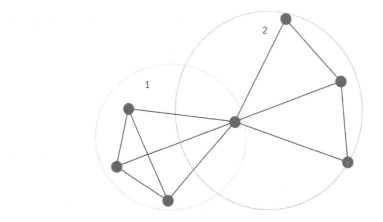

Fig. 1. A sample network

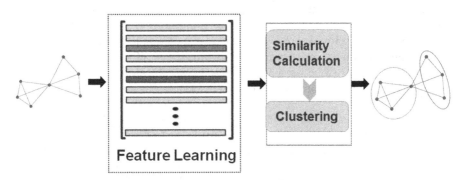

Fig. 2. The architecture of newly proposed machine learning methods.

e.g. cosine similarity. The last step is traditional clustering which is based on the calculated similarities. For this architecture, it is important how to learn representations of the given networks, because network representations affect the performance of the following community detection progress. In this paper, we propose a new method based on edge representation learning which can not only detect overlapping communities but also improve the performance of community detection. We sample a series of edge sequences by random walks on graph, then utilize unsupervised machine learning method to learn a mapping from edges to feature representations. After the mapping progress, we divide edges into groups by traditional clustering method, e.g. k-means algorithm [8] and its variants. Each edge connects two nodes, that is to say, one edge corresponds two nodes. In this way, we transform edges communities into corresponding nodes communities. As a result, the fuzzy nodes in networks that belong to different communities simultaneously can be detected. Our contributions are:

(1) We propose a scalable edge embedding method to learn edge representations based on language modeling tools.

(2) We apply edge representation learning method to community detection, and propose a new way to find overlapping communities by clustering edges.

(3) We conduct multiple experiments on synthetic networks and real world networks to test the performance of proposed algorithm.

The rest of the paper is organized as follows. Section 2 overviews related work in community detection domain. Section 3 introduces edge representation learning model. In Sect. 4, we present steps on how to perform CD-ERL (community detection based on edge representation learning) algorithm. Section 5 displays experiments and results. Section 6 concludes the paper.

2   Related Work

In the past few years, a lot of community detection algorithms have been proposed [11,24]. Even though they are both for finding community structure, the principles are different. Some algorithms are based on modularity optimization, such as FN (Fast-Newman) algorithm [17]; some are based on graph partitioning, such as KL (Kernighan-Lin) algorithm [9]; there are also some algorithms which are based on label propagation [19], spectral clustering [16] and dynamic networks [2]. However most of them just fit networks with clear structure.

With the rapid development of representative learning for natural language processing, there are some new representative methods proposed in network field [1,18,21,23], such as node2vec [7]. These methods provide a new guidance for network analysis. However in community detection area, these methods that divide nodes directly can only find non-overlapping communities, while communities in real world networks are always overlapping. They ignore the importance of the nodes that belong to more than one communities.

Inspired by newly proposed representative learning methods, we proposed a CD-ERL algorithm to detect overlapping communities by clustering edges. The same way as representative learning methods for natural language processing, we sample edge sequence according to network links, and learn a mapping from edges to vectors. However, there are many representative learning methods used in community detection area, they only consider nodes representation and nodes clustering which lead to non-overlapping communities. We overcome this by learning edge representations, and by clustering edges, we can get both edge communities and corresponding node communities. The node communities we detect are overlapping, and the fuzzy nodes can be detected.

3   Edge Representation Learning Model

In this paper we formulate the edge representation learning progress as a maximum likelihood problem by extending representative learning for natural language processing into networks. Feature representation learning methods based on Skip-gram architecture [13] have been originally developed in the context of natural language. Here we use Skip-gram model as an example to do community detection task.

Let G = (V, E) be a given network. Here V is the set of nodes, E is the set of edges in the network. We assume G is an undirected, unweighted network here. Our goal is to embed edges in E into d-dimensional feature space R^d. Let $f : E \to R^d$ be the mapping function from edges to feature matrix, while $R^d(e)$ is the feature representation of edge e. We regard edges in a graph as words in a document, and edge neighbors as words before and after the target word. For e in E, let $N(e)$ represents the neighbors of edge e. In the embedding progress, we want to preserve the neighbor relations as much as possible. Thus the objective function is as follows:

$$\max \sum_{e \in E} \log Pr(N(e)|R^d(e)) \tag{1}$$

We assume that given the feature representation of an edge, finding one neighbor edge is independent of finding another neighbor edge. What's more, interactive edges have symmetry effect on each other. We use a softmax unit to model the conditional likelihood of the source edge and its neighbor. In these conditions, the objective function can be written in this form:

$$\max \sum_{e \in E} [-\log \sum_{e_1 \in E} exp(R^d(e) \cdot R^d(e_1)) + \sum_{i \in N(e)} R^d(i) \cdot R^d(e)] \tag{2}$$

Then we optimize it using stochastic gradient ascent over the model parameters. The detailed steps of edge representation learning algorithm can be summarized in Algorithm 1.

Algorithm 1. Edge Representation Learning Algorithm

Input: Graph(V, E), Dimensions d, Walks per edge r, Walk length l, Context size k
1: Initialize walks to Empty
2: **for all** $iter = 1$ to r **do**
3:     **for all** edges $e \in E$ **do**
4:         Walk = RandomWalk(G, e, l)
5:         Append walk to walks
6:         f = StochasticGradientDescent $(k, d, walks)$
7:         **return** f
8:     **end for**
9: **end for**

---

4   CD-ERL Algorithm

In this section, the details of CD-ERL algorithm are described. There are four steps to complete CD-ERL algorithm.

Step 1: **Edge sequences sampling.** Edge sequence in network is non-linear, unlike word sequence in document. It is important how to sample edge sequences. Given a pronounced community structure, we want to achieve this effect that

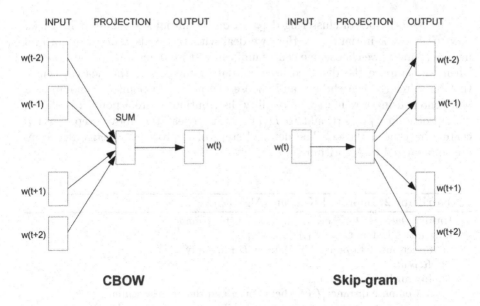

Fig. 3. The architectures of CBOW and Skip-gram model. [13]

pairs of edges in the same community are much more easily reachable by the sampling strategy than pairs of edges in different communities. In this paper, we choose random walk as the sampling strategy since random walk can well meet our requirements. We collect edge neighbors by random walk, and regard the sampled edge sequences in a network as word sequences in a document.

Step 2: **Edge representations learning.** After getting sampled edge sequences, based on representative learning method for natural language processing, e.g. continuous bag of words [14] and Skip-gram model [13], we transform edges in graph into representative feature matrix R^d using an unsupervised machine learning way. CBOW and Skip-gram model are two widely used methods in natural language processing field. As shown in Fig. 3, the CBOW architecture predicts the current word based on the context, and the Skip-gram predicts surrounding words given the current word. At the same time the representation of the inputs are automatically learned. Here we replace the words in the context by sampled edge sequences.

Step 3: **Edges clustering.** In this part, we choose an improved k-means algorithm to complete edge clustering. Given an edge feature matrix R^d, the objective function is:

$$E = \sum_{i=1}^{k} \sum_{x \in C_i} ||x - u_i||_2^2 \qquad (3)$$

Our goal is to minimize it. Here u_i is the mean vector of cluster C_i. K-means algorithm [8] adopts greedy strategy and gets the approximate solution by iterative optimization.

The performance of clustering depends on the initial k seeds, so it is crucial how to choose k initial seeds. Here we deal with the seeds using an improved method. Firstly, we choose one center uniformly at random from all data points. Then we compute the distance between data point x and the nearest center that has already been chosen, and choose one new data point at random as a new center, using a weighted probability distribution where a point x is chosen with probability proportional to $D(x)^2$. Last, repeat the above steps until k centers have been chosen. The detailed steps of improved K-means algorithm are summarized in Algorithm 2.

Algorithm 2. Improved K-means Algorithm

Input: Dataset $D = \{x_1, x_2, ..., x_m\}$, Cluster Number k
Output: Clusters $C = \{C_1, C_2, ..., C_k\}$
1: Select one data point u in Dataset D randomly
2: *Repeat*
3: **for all** $l = 1$ to m **do**
4:     Compute distance $D(x_l)$ between x_l and the nearest center
5:     Choose a new center using a probability proportional to $D(x_l)^2$
6: **end for**
7: Until k centers have been chosen
8: *Repeat*
9: Let $C_i = \phi$ $(1 \le i \le k)$
10: **for all** $j = 1$ to m **do**
11:     Compute $d_{ji} = ||x_j - u_i||_2$
12:     $\lambda_j = argmin_{i \in 1,2,...,k} d_{ij}$
13:     $C_{\lambda_j} = C_{\lambda_j} \sqcup x_j$
14: **end for**
15: **for all** $i = 1$ to k **do**
16:     Compute the new mean vectors : $u_{i1} = \frac{1}{|C_i|} \sum_{x \in C_i} x$
17:     **if** $u_{i1} \ne u_i$ **then**
18:         Update u_i to u_{i1}
19:     **else**
20:         Keep the same u_i
21:     **end if**
22: **end for**
23: Until all the mean vectors do not change

Step 4: **Transforming edge communities into node communities.** In graphs, each edge connects two nodes, in other words, one edge corresponds two nodes. In this way, we can transform the edge communities into corresponding node communities. In the example above, the edges connecting A with B and A with C would be placed in different groups, and since they both have A as endpoint, the latter turns out to be an overlapping vertex. Thus nodes belongs to more than one communities can be detected, and we can get both edge communities and node communities.

The complete pseudocode of CD-ERL algorithm is Algorithm 3.

Algorithm 3. CD-ERL Algorithm

Input: A network $G(V, E)$, edge representation learning parameters, number of
    communities k
1: vectors = ERL(ERL parameters)
2: edge communities = K-means(vectors, k)
3: turn edge communities into node communities
4: return node communities

---

5 Experiments and Results

To better explain the process of the proposed CD-ERL algorithm, we test it on
Zachary's karate club dataset [25].

5.1 A Running Instance

Zachary's karate club dataset is widely used in network analysis field. A social
network of a karate club was studied by Wayne W. Zachary for a period of three
years from 1970 to 1972. The network captures 34 members of a karate club,
documenting 78 pairwise links between members who interacted outside the club.
The visualization of this network is shown in Fig. 4. In this part, the parameters
used in CD-ERL algorithm is set as Table 1. We extract 5 walks per edge from

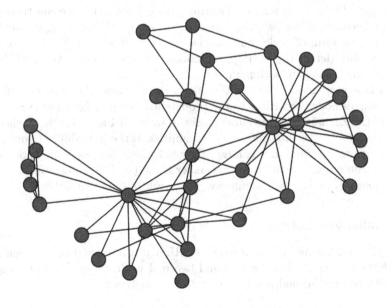

Fig. 4. Zachary's karate club dataset.

Table 1. Experiment parameters

Parameters	# walks	Walk length	Dimensions	Communities
Zachary's karate club networks	5	10	8	2

Table 2. Edge communities

	Edge
Community 1	1,3,4,5,6,8,10,11,12,14,15,16,17,20,21,24,26,28,29,31,32,33,34,35,37,39, 40,41,42,43,46,48,49,50,51,52,54,55,56,59,61,63,64,65,67,68,69,73,76
Community 2	0,2,7,9,13,18,19,22,23,25,27,30,36,38,44,45,47,53,57,58,60,62,66,70,71, 72,74,75,77

Table 3. Node communities

Node	0	1	2	3	4	...	9	10	11	12	...
Community 1	0.6875	0.4444	0.7000	0.8333	0.6667	...	0.5000	0.6667	0	1	...
Community 2	0.3125	0.5556	0.3000	0.1667	0.3333	...	0.5000	0.3333	1	0	...

the original network. Every walk is an edge sequence, and the length of each edge sequence is 10. We input these extracted walks into the representation learning model and set the feature representation dimension, then an edge representation will be automatically learned.

The community detection result of CD-ERL algorithm is shown in Table 2, it is the detected edge communities. Getting edge communities, we can transform it into node communities based on the correlation of nodes and edges. The final result with the form of node communities is shown in Table 3. We can see that it can not only detect the overlapping nodes, but also provide the probability that one node belongs to a community.

This is a very good property because we can decide the final community membership of nodes according to the probabilities we get. For example, for the results of the Zachary's karate club dataset, if we set 0.4 as a threshold, then we can divide these nodes into different communities. If the probability of one node that belongs to a community is larger than 0.4, then we believe this node is in the above community, if the value is smaller than 0.4 then we believe this node can't belongs to the above community. The result is shown in Table 4.

5.2 Evaluation Metrics

To test the performance of proposed CD-ERL algorithm, we do experiments on nine synthetic network benchmarks and two real world networks, and compare them with two traditional community detection methods.

Table 4. Community detection results

Node ID	Community 1	Community 2
0	YES	NO
1	YES	YES
2	YES	NO
3	YES	NO
4	YES	NO
...	...	...
9	YES	YES
10	YES	NO
11	NO	YES
12	YES	NO
...	...	...

NMI. In this paper, we use NMI (Normalized Mutual Information) to evaluate the performance of community detection algorithms. The definition of NMI is as follows:

$$\text{NMI} = \frac{2I(X;Y)}{H(X) + H(Y)} \tag{4}$$

If there are N samples in both X and Y, N_i is the number of samples which is equal to i, N_j is the number of samples which is equal to j, N_{ij} is the number of samples which is equal to i in X, equal to j in Y. The calculation formula of NMI [4,22] becomes as follows:

$$\text{NMI} = \frac{-2\sum_{ij} N_{ij} \log \frac{N_{ij} \cdot N}{N_i \cdot N_j}}{\sum_i N_i \cdot \log \frac{N_i}{N} + \sum_j N_j \cdot \log \frac{N_j}{N}} \tag{5}$$

Its lower bound is 0, representing the independence of the result and ground truth, and its upper bound is 1, representing community detection result is the same with ground truth. The closer NMI score is to 1, the better the community detection result is.

V-measure. A clustering result satisfies homogeneity if all of its clusters contain only data points which are members of a single class. A clustering result satisfies completeness if all the data points that are members of a given class are elements of the same cluster. V-measure [20] is the harmonic mean between homogeneity and completeness.

$$V = \frac{2 \cdot homogeneity \cdot completeness}{homogeneity + completeness} \tag{6}$$

It is symmetric. Its bound is between 0 to 1. 1 stands for perfectly complete community detection. The closer V-measure score is to 1, the better the community detection result is.

Since CD-ERL algorithm can find overlapping communities, in order to perform a better contrast with the traditional non-overlapping community detection algorithms, we make a preprocessing for the results of CD-ERL algorithm. If nodes belongs to more than one communities, we will randomly set this node to one of these communities. In this way, we can contrast the CD-ERL algorithm with traditional non-overlapping community detection methods.

5.3   Baseline Methods

In this paper, we employ two traditional representative community detection algorithms, the LPA (Label Propagation Algorithm) and Louvain algorithm, as the contrast with our method.

Label Propagation Algorithm. This algorithm was introduced by Raghavan et al. [19]. It proceeds based on the assumption that each node in the network is in the same community with the majority of its neighbors. Every node is initialized a distinct label at the start. In the process of iteration, for a node, count its neighbors' labels and label the node according to the majority of its neighbors' labels. When each node in network has the same label as the majority of its neighbors, the iteration stops. The LPA algorithm is suitable for large scale networks.

Louvain Algorithm. The Louvain community detection algorithm [3] optimizes the modularity of a partition of the network by greedy optimization. It optimizes local modularity and generates small scale communities. Then it regards the small communities as nodes to generate a new network. These steps are repeated iteratively until a maximum of modularity is attained and a hierarchy of communities is produced. Relatively speaking, this method is fast and accurate.

5.4   Dataset

Synthetic Benchmark Networks. LFR(Lancichinetti-Fortunato-Radicchi) benchmark graph [10] is one of the most frequently-used synthetic network generating model in community detection filed. This method introduces power-law distributions of degree and community size to the graphs to simulate real world networks. There are six main parameters in this model. N is the number of nodes, k is the average degree, $maxk$ is the maximum degree of nodes, $minc$ is the minimum number of community members, $maxc$ is the maximum number of community members. μ is mix parameter, which is defined as:

$$\mu = \frac{\sum_i k_i^{ext}}{\sum_i k_i^{tot}} \tag{7}$$

Here k_i^{ext} and k_i^{tot} stand for the external degree of node i, i.e. the number of edges connecting it to others that belong to different communities, and the total degree of said node. In this paper, we generate nine networks with different mu. Since the definition of community that in the same communities nodes are linked densely, between different communities the links are sparse, we choose μ from 0.1 to 0.5 for the following experiments. The settings of parameters for the synthetic networks are shown in Table 5.

Table 5. Parameter settings of synthetic networks

No.	N	k	maxk	minc	maxc	μ
1	10000	8	20	3	1000	0.10
2	10000	8	20	3	1000	0.15
3	10000	8	20	3	1000	0.20
4	10000	8	20	3	1000	0.25
5	10000	8	20	3	1000	0.30
6	10000	8	20	3	1000	0.35
7	10000	8	20	3	1000	0.40
8	10000	8	20	3	1000	0.45
9	10000	8	20	3	1000	0.50

Real World Networks. In this paper, we also evaluate our algorithm on two real world networks. These networks with ground truth communities are both from SNAP(Stanford Network Analysis Platform) [12]. We conduct contrast experiments on Amazon and Youtube networks. These two datasets contain not only the graph links, but also the ground truth communities.

We observe that there are some stray nodes with few links in networks which will damage the community detection progress, so we treat them as noisy nodes. Thus, we preprocess these datasets by deleting these stray nodes. After that we get two high-quality networks whose properties are shown in Table 6.

Table 6. Properties of real world networks

Name	Type	Nodes	Edges	Average degree	Description
Youtube	Undirected & unweighted	39841	224235	11.2565	Youtube online social network
Amazon	Undirected & unweighted	16716	48739	5.8314	Amazon product network

5.5 Result Analysis

To test the performance, we run CD-ERL algorithm and the contrast methods on both synthetic and real world networks. The parameters used in CD-ERL algorithm is set as Table 7.

Table 7. Experiment parameters

Parameters	# walks	Walk length	Dimensions	Communities
Synthetic networks	20	40	128	1000
Real world networks	20	40	128	5000

We generate nine different synthetic networks with different values of μ from 0.1 to 0.5. We repeat experiments for five times on every synthetic networks, and regard the average results as the final results. Experiment results on synthetic benchmark networks are shown in Figs. 5 and 6.

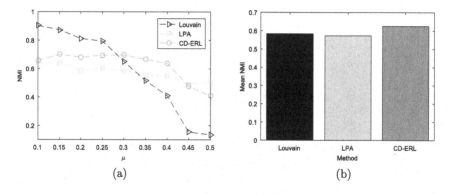

Fig. 5. A contrast of NMI using different methods on networks with different μ.

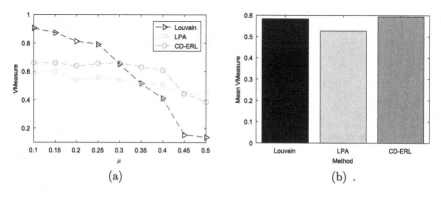

Fig. 6. A contrast of V-measure using different methods on networks with different μ.

In Figs. 5 and 6, horizontal axis represents different mu, while vertical axis represents the value of NMI and V-measure score. We can see that there is a rapid decrease for Louvain algorithm when mu is getting larger. On the contrast, CD-ERL and LPA algorithm is stable with the increase of mu. But the CD-ERL algorithm that we propose in this paper performs better than LPA. For mu from 0.1 to 0.25, Louvain algorithm gets higher value of NMI than the others. The reason is when mu is small, networks usually have clear structure, thus Louvain algorithm using modularity optimizing strategy deal with this kind of networks better. But real world networks are always complex and uncertain, Louvain algorithm cannot solve the problems stably. CD-ERL and LPA are stable in networks with different structure. But we can see that for the mean NMI and V-measure score, CD-ERL performs best in contrast with other two methods. When dealing with real world networks with complex and fuzzy structure, CD-ERL is more suitable.

What's more, we also employ these algorithms on real world networks. Experiment results on real world networks are shown in Figs. 7 and 8.

It is obvious in Figs. 7 and 8 that the performance of CD-ERL algorithm is much better than LPA and Louvain algorithm. The main idea of Louvain algorithm is to maximize the modularity of networks, it is suitable for networks with obvious community structure. Thus, when the real world network is fuzzy, Louvain algorithm gets bad results. The performance of LPA algorithm is between Louvain algorithm and ours. For LPA algorithm, the label of node in network

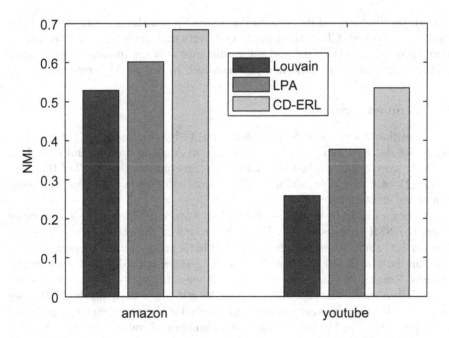

Fig. 7. A contrast of NMI using different methods on different real world networks.

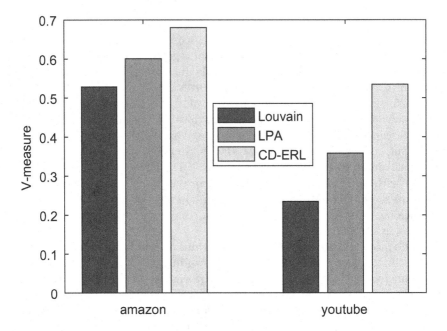

Fig. 8. A contrast of V-measure using different methods on different real world networks.

depends on the labels of its surrounding nodes. When the network is fuzzy, the result becomes bad. CD-ERL algorithm converts the graph links into vectors by edge representation learning, and it uses distance measure instead of modularity in clustering process, this method performs well in real world networks.

5.6 Parameter Sensitivity

In this section, we discuss the parameter sensitivity of proposed CD-ERL algorithm. We do the community detection task on Amazon dataset and show how the NMI score changes when we change the three parameters of CD-ERL algorithm: (1) walk length of random walk (2) number of walks per edge (3) dimension of the embedding.

As shown in Fig. 9, we can see that when feature dimension is set to a small value, the NMI score is higher. With the growth of dimension, the NMI score drops a bit, then keeps a steady state. What's more, we do experiments on Amazon dataset to see how the NMI score changes with different walk length settings. Results in Fig. 10 show when walk length is set to 20, the NMI score is low. When we do random walks of walk length larger than 40, the NMI score increases a lot and remains constant. Similarly for the parameter of number of walks per edge, Fig. 11 shows with a small number of walks per edge, we get a low NMI score. But when the number of walks is larger than 30, the NMI score increases a bit. And with the growth of number of walks per edge, the NMI score

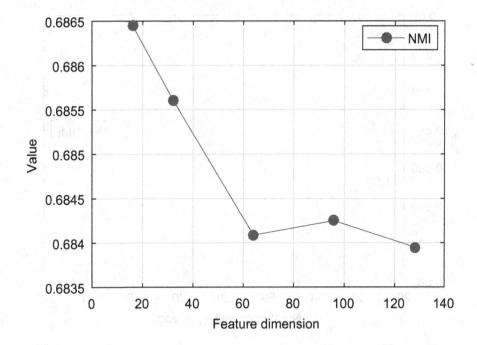

Fig. 9. NMI score on Amazon dataset for various values of dimensions.

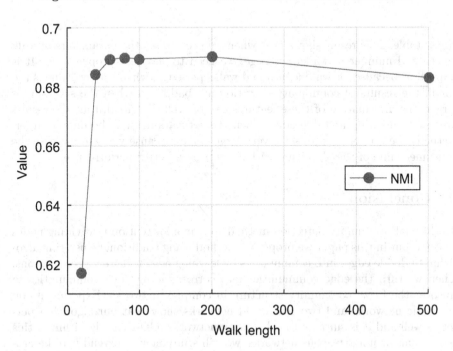

Fig. 10. NMI score on Amazon dataset for various values of walk length.

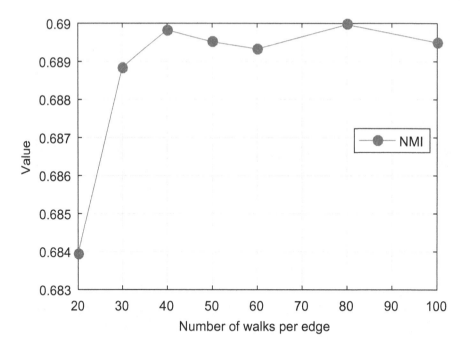

Fig. 11. NMI score on Amazon dataset for various values of number of walks per edge.

stays stable. This result shows that when we need to set the parameters of walk length and number of walks per edge, we need to choose a proper value. It is important because for walk length and walks per edge, when the set value is too small, the results of community detection are bad. And when the set value is larger, the information of these features can be well described, thus the results are good and keep a stable state when the set parameter is becoming larger. Actually, we do not need to set a very large value, because you will increase the consumed time of the algorithm without getting a better performance.

6   Conclusion

Traditional community detection methods pay more attention to dividing nodes directly, but in this paper we propose a method using traditional clustering algorithm to divide edges into communities by learning edge feature representations. Then we turn the edge communities into corresponding node communities to uncover the fuzzy community structure in complex networks. Experiments on synthetic networks and two real world networks show that our algorithm performs well and it is suitable to solve fuzzy networks. CD-ERL algorithm in this paper aims at homogeneous networks, we will study how to extend it to heterogeneous networks next.

Acknowledgments. The work presented in this paper was supported in part by the Special Project for Independent Innovation and Achievement Transformation of Shandong Province (2013ZHZX2C0102, 2014ZZCX03401).

References

1. Adhikari, B., Zhang, Y., Ramakrishnan, N., Prakash, B.A.: Distributed representation of subgraphs. arXiv preprint arXiv:1702.06921 (2017)
2. Bansal, S., Bhowmick, S., Paymal, P.: Fast community detection for dynamic complex networks. In: da F. Costa, L., Evsukoff, A., Mangioni, G., Menezes, R. (eds.) CompleNet 2010. CCIS, vol. 116, pp. 196–207. Springer, Heidelberg (2011). https://doi.org/10.1007/978-3-642-25501-4_20
3. Blondel, V.D., Guillaume, J.L., Lambiotte, R., Lefebvre, E.: Fast unfolding of communities in large networks. J. Stat. Mech. Theory Exp. **2008**(10), P10008 (2008)
4. Danon, L., Diaz-Guilera, A., Duch, J., Arenas, A.: Comparing community structure identification. J. Stat. Mech. Theory Exp. **2005**(09), P09008 (2005)
5. Fortunato, S.: Community detection in graphs. Phys. Rep. **486**(3), 75–174 (2010)
6. Fortunato, S., Hric, D.: Community detection in networks: a user guide. Phys. Rep. **659**, 1–44 (2016)
7. Grover, A., Leskovec, J.: node2vec: scalable feature learning for networks. In: Proceedings of the 22nd ACM SIGKDD International Conference on Knowledge Discovery and Data Mining, pp. 855–864. ACM (2016)
8. Jain, A.K.: Data clustering: 50 years beyond k-means. Pattern Recogn. Lett. **31**(8), 651–666 (2010)
9. Kernighan, B.W., Lin, S.: An efficient heuristic procedure for partitioning graphs. Bell Syst. Tech. J. **49**(2), 291–307 (1970)
10. Lancichinetti, A., Fortunato, S., Radicchi, F.: Benchmark graphs for testing community detection algorithms. Phys. Rev. E **78**(4), 046110 (2008)
11. Leal, T.P., Goncalves, A.C., Vieira, V.d.F., Xavier, C.R.: Decode-differential evolution algorithm for community detection. In: 2013 IEEE International Conference on Systems, Man, and Cybernetics (SMC), pp. 4635–4640. IEEE (2013)
12. Leskovec, J., Krevl, A.: SNAP Datasets: Stanford large network dataset collection, June 2014. http://snap.stanford.edu/data
13. Mikolov, T., Chen, K., Corrado, G., Dean, J.: Efficient estimation of word representations in vector space. arXiv preprint arXiv:1301.3781 (2013)
14. Mikolov, T., Sutskever, I., Chen, K., Corrado, G.S., Dean, J.: Distributed representations of words and phrases and their compositionality. In: Advances in Neural Information Processing Systems, pp. 3111–3119 (2013)
15. Newman, M.E.: The structure and function of complex networks. SIAM Rev. **45**(2), 167–256 (2003)
16. Newman, M.E.: Finding community structure in networks using the eigenvectors of matrices. Phys. Rev. E **74**(3), 036104 (2006)
17. Newman, M.E.: Modularity and community structure in networks. Proc. Nat. Acad. Sci. **103**(23), 8577–8582 (2006)
18. Perozzi, B., Al-Rfou, R., Skiena, S.: Deepwalk: online learning of social representations. In: Proceedings of the 20th ACM SIGKDD International Conference on Knowledge Discovery and Data Mining, pp. 701–710. ACM (2014)
19. Raghavan, U.N., Albert, R., Kumara, S.: Near linear time algorithm to detect community structures in large-scale networks. Phys. Rev. E **76**(3), 036106 (2007)

20. Rosenberg, A., Hirschberg, J.: V-measure: a conditional entropy-based external cluster evaluation measure. EMNLP-CoNLL **7**, 410–420 (2007)
21. Tang, J., Qu, M., Wang, M., Zhang, M., Yan, J., Mei, Q.: Line: large-scale information network embedding. In: Proceedings of the 24th International Conference on World Wide Web, pp. 1067–1077. ACM (2015)
22. Vinh, N.X., Epps, J., Bailey, J.: Information theoretic measures for clusterings comparison: is a correction for chance necessary? In: Proceedings of the 26th Annual International Conference on Machine Learning, pp. 1073–1080. ACM (2009)
23. Wang, S., Tang, J., Aggarwal, C., Chang, Y., Liu, H.: Signed network embedding in social media. In: SDM (2017)
24. Yang, L., Cao, X., Jin, D., Wang, X., Meng, D.: A unified semi-supervised community detection framework using latent space graph regularization. IEEE Trans. Cybern. **45**(11), 2585–2598 (2015)
25. Zachary, W.W.: An information flow model for conflict and fission in small groups. J. Anthropol. Res. **33**(4), 452–473 (1977)

Introducing ADegree: Anonymisation of Social Networks Through Constraint Programming

Sergei Solonets, Victor Drobny, Victor Rivera[✉], and JooYoung Lee

Institute of Technologies and Software Development,
Innopolis University, Innopolis, Russia
{s.solonets,v.drobnyy,v.rivera,j.lee}@innopolis.ru

Abstract. With the rapid growth of Online Social Networks (OSNs) and the information involved in them, research studies concerning OSNs, as well as the foundation of businesses, have become popular. Privacy on OSNs is typically protected by anonymisation methods. Current methods are not sufficient to ensure privacy and they impose restrictions on the network making it not suitable for research studies. This paper introduces an approach to find an optimal anonymous graph under user-defined metrics using Constraint Programming, a technique that provides well-tested and optimised engine for combinatorial problems. The approach finds a good trade-off between protection of sensitive data and quality of the information represented by the network.

Keywords: Anonymisation · Online social networks
Constraint Programming

1 Introduction

Social networks (SNs) are social structures made up of individuals (or organisations) that are connected by one or more types of interdependency. Individuals are called "nodes" and connections are called "edges". There are several types of SNs; for example, in LinkedIn[1] (an online network of professionals) every link between two users specifies a professional relationship between them [10]. In Facebook[2] or VK[3] links correspond to friendship. Each SN shares information according to the type of the network. Given the rapid growing of SN and the information involved in the networks, several research studies have been made [7,9,11,12,14,16].

The information carried out by Social Networks is of paramount importance in domains such as marketing [8]. Network owners often share this information with advertising partners and other third parties. Such practice is the foundation of the business case for many social network owners, that have made the analysis of social networks profitable. However, these kind of marketing makes

---

[1] www.linkedin.com.
[2] www.facebook.com.
[3] www.vk.com.

C. Doulkeridis et al. (Eds.): MATES 2017, LNCS 10731, pp. 73–86, 2018.
https://doi.org/10.1007/978-3-319-73521-4_5

the owners of social networks like `Facebook` and `Twitter` (http://www.twitter.
com) increase the sharing of potentially sensitive information about users and
their relationships. Owners are compelled by law to respect the privacy policies
of their users. This privacy is typically protected by anonymisation.

Anonymisation methods are used to assure privacy policies of the users of
SN (i.e., sensitive information). The common intuitive method is to remove the
identity of each node in the graph, replacing it with a random identification
number. However, Backstrom et al. [1] have shown that this method is not
adequate for preserving the privacy of nodes. Specifically, the authors show that
in such an anonymised network, there exists an adversary who can identify target
individuals and the link structure between them.

There have been some other attempts to come up with more advanced
anonymisation approaches. As an example, Liu and Terzi [13] consider node
re-identification assuming that the adversaries' auxiliary information consists
only of node degrees. Campan and Truta [3] propose metrics for the information loss caused by edge addition and deletion and apply k-anonymity to node
attributes as well as neighbourhood structure, among others [6,20].

Despite the fact that there exist many anonymisation methods, those concerning to remove private information of nodes and those that change the structure of the network (e.g., based on k-degree anonymity) are not sufficient for
privacy when dealing with social networks [1,15]: the fundamental issue when
removing private information of nodes (e.g. names, addresses) is that even though
there are millions of people participating in a social network, each node has a
relatively unique relationship with his/her neighbours. This uniqueness can be
exploited in order to identify the participants in the network; the fundamental issue when changing the structure of the graph (e.g. based on k-anonymity)
is that they impose arbitrary restrictions on the network and make arbitrary
assumptions about the properties of the given graph [15].

We consider a scenario where the owner of a social network wants to release
the underlying SN graph preserving the privacy of its users. This paper introduces the formal definition of an anonymisation approach using Constraint Programming (CP) in its implementation. Our approach finds out a network that
respects the privacy polices, hides the private information, makes it harder to
de-anonymise the network and, as opposed to current anonymisation approaches,
conserves the properties of the network as much as possible. On the other hand,
[13] only considers the degree attack scenario. Due to the nature of information
carried out by the Social Networks, only k-automorphism constraint can guarantee privacy under any type of attack [21]. However this kind of constraint is
very heavy and as a consequence leads to a huge information loss. Therefore,
to find a network as close as possible to the original, we need to study possible
scenario attacks in every particular case.

2 Preliminaries

2.1 Constraint Programming

Constraint programming (CP) [17] is a paradigm for solving combinatorial search problems that draws on a wide range of techniques. CP is currently applied with success to many domains, such as scheduling, planning, vehicle routing, configuration, networks, and bioinformatics. The focus of CP is on reducing the search space by pruning values that cannot appear in any feasible or optimal solution. Constraints are relations, and a constraint satisfaction problem (CSP) states which relations should hold among the given decision variables. Constraint solvers take a real-world problem and represent in terms of decision variables and constraints to find an assignment to all the variables that satisfies the constraints. To get a solution of a CSP, one uses the following generic algorithm:

1. Use propagation to prune the domains of the variables. This step is executed by propagators. Propagators are processes that filter the domains of a set of finite domain variables according to the semantics of the constraint they implement (also called pruning rules). Propagators share a common store which contains the information that is currently known about the variables of the problem. As soon as a propagator is able to infer new information from the store, it adds this new information in the store. This iterative steps stop when the propagator(s) reaches a fixed point, i.e., when it cannot prune more values.
2. When the propagators reach a fix point, we may have three possible situations:
   - all the variables are bound to a value. In this case the search stops since a solution has been obtained.
   - there are some variables that are yet to be determined. In this case, we split the CSP into new problems according to branch strategies, generating new constraints. Then the step 1 is triggered, we wait until propagators get stable to continue.
   - there is at least one variable whose domain is empty. In this case, we go back to the previous step.

2.2 Social Networks as Graphs

A social network can be seen as a social structure represented by a graph where nodes represent actors of the social network and edges represent a interdependency among those actors. Edges can represent specific types of interdependency, such as friendship, kinship, common interest, financial exchange, dislike, and so forth. The following is the definition of a Social Network which will be used throughout this paper.

Definition 1 (Social Network). *A social network SN is a directed/undirected graph* $G = (V, E)$, *containing*

1. *a set of attributes for each node in V and*
2. *a set of attributes for each edge in E.*

Typically, Social Network analysers are interested in sub-populations of social graphs or sub-links of the graphs. For instance, demographic studies might be interested in relationships between people of the same gender in ages of a specific range. This categorization of graphs is represented by Restricted Graphs:

Definition 2. *Restricted Graph:* *Let* $G(V, E)$ *be a graph.* $G^- \langle V^-, E^- \rangle$ *is a restricted graph with* $V^- \subseteq V$ *and* $E^- \subseteq E$

An interesting property of data anonymisation is k-anonymity introduced in [19]. k-anonymity solves the problem: "Given person-specific field-structured data, produce a release of the data with scientific guarantees that the individuals who are the subjects of the data cannot be re-identified while the data remain practically useful." Liu and Terzi [13] introduced the concept of k-degree anonymous graphs. The following is the formal definition as used in the paper. We first define a degree sequence for the graph.

Proposition 1. *Degree Sequence:* *Let* dG *be sequence.* dG *is a degree sequence of graph* $G(V, E)$ *iff is a sequence of size* $n = |V|$ *such that* $dG[i]$ *is the degree of the i-th node of* G.

The definition of k-anonymity for the degree sequence is as follows.

Proposition 2. k-*degree Anonymous:* *A degree sequence* dG *is* k-*anonymous, if every distinct value in* dG *appears at least* $k - 1$ *times. That is*

$$\forall i \in dG \Rightarrow occur \ (i, dG) \geq k,$$

where $occur \ (v, s)$ *is the number of occurrences of* v *in sequence* s.

Finally, we state that

Definition 3. *a* k-*degree anonymous graph has associated a* k-*degree anonymous sequence.*

3   Related Work

Privacy on social networks is typically protected by anonymisation methods. An initial anonymisation attempt was to remove sensitive information of the nodes, such as names, addresses and phone numbers. However, this kind of method is easy to de-anonymise by studying the structure of the network. There have been several anonymisation methods that modify the initial structure of graphs that represent Social Networks to avoid users being uncovered by analysing the structure of their connections. Liu and Terzi anonymise the network by making

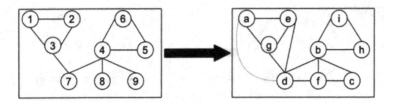

Fig. 1. Liu et al. approach.

it k-degree anonymous [13]. A network is k-degree anonymous if for every node v, there exist at least $k-1$ other nodes in the network with the same degree as v. Figure 1 shows an example of such anonymisation.

Campan and Truta suggest anonymising the network by applying the concept of edge generalizing on the corresponding graph [3]. Edge generalisation is implemented by clustering nodes. Every cluster is replaced with a new node whose neighbours are the union of the neighbours of the nodes in the cluster. Each new node is associated with a pair of integers (#n, #e) representing the number of nodes and the number of edges inside the cluster. These pairs of integers are then used to approximate the interconnection of those nodes inside the cluster without revealing information about how two nodes in separate clusters are interconnected. Figure 2 shows an example of such anonymisation. Notice that in each case the interconnection of the nodes inside every cluster can be inferred. For instances, there is only one instance of connecting 3 nodes with 3 edges. However, the connections between nodes of different clusters are hidden by the anonymisation process.

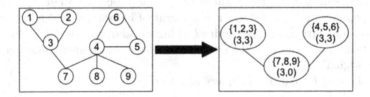

Fig. 2. Campan et al. approach.

Bhagat et al. anonymise the network by associating each node with a list of possible *ids* [2]. Each list contains the real *id* of the node. The sensitive information is therefore protected, since, in general, it is not possible to infer whether a node is present in the network and whether there is a connection between two nodes. Bhagat et al. shows, however, that these lists of *ids* need to be carefully computed since it is not impossible to unmask individuals and connections. Figure 3 shows an example of such anonymisation. Notice that in every case the *id* of the node has been replaced with a list of possible *ids* containing the real one.

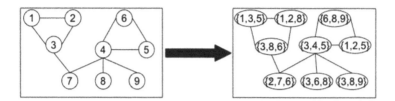

Fig. 3. Bhagat et al. approach.

4   Anonymising Social Network: ADegree

ADegree is a constraint that is used to find an anonymised network that protects
users' sensitive information by ensuring k-degree anonymity, making the reverse
process (de-anonymisation) harder to achieve. The anonymised network pre-
serves the structure of the original network as much as possible. This structure
is then preserved by defining the interest of the Social Network's analyser. Typically, Social Networks analysers are interested in subgraphs within the Social
Network. Section 4.1 defines a way to represent such interest and Sect. 4.2 formally defines the constraint for the anonymisation of Social Networks.

4.1   Query of Interest

The fundamental problem when altering the structure of a graph is that arbitrary restrictions are imposed on the network as well as arbitrary assumptions
are made about its properties. The problem is due to the ignorance on the utility
of anonymised graph. Companies dedicated to the study of social networks have
specific queries of interest. For example, 'how many users are there in some specific sub-populations?', 'What are the patterns of interaction and friendships?',
'Which sub-populations are interacting?'. These kind of queries define the utility
of the network and can be used to determine how close an anonymised graph is
from the original one.

The following formalises the query of interest.

Definition 4. *Query of Interest:* *A query of interest QI is a tuple $\langle G, P,$
$T, F \rangle$ where*

$G(V, E)$ is a graph (representing a Social Network);
$P \subseteq \mathbb{P}(V)$ is set of set of nodes;
$T \subseteq \mathbb{P}(label(E))$ is set of set of labels.;
$F \in \{G^- \langle p, t \rangle \mid p \in P \wedge t \in T\} \to \mathbb{Z}$
where $\mathbb{P}$ is the power set of a set.

In Definition 4, P defines the different categories of 'users' (actors) that the
query is interested in. T defines the different categories of 'labels' that the query
is interested in. As an example, take the phone communication company AT&T
social network that represent people as nodes, phone calls between people as

edges and the duration of calls as edges' labels. In a possible QI, P categorises the set of user the company is interested in, e.g. teenagers and females over 50 years, and T categorises the set of labels, e.g. 3 to 15 min. F is a function (defined by the user) that represents the closeness of G^- to G.

4.2 ADegree

The constraint is defined as follows.

Definition 5. *ADegree: Suppose $G(V, E)$ and $G'(V, E')$ are graphs representing social networks. Let $QI = \langle G, P, T, F \rangle$ be the query of interest, and let k and v be integers. The constraint ADegree holds iff*

1. *G' is k-degree anonymous;*
2. *G' is as close as possible to G*

$$\sum_{p \in P, t \in T} abs(F(G, p, t) - F(G', p, t)) \leq v;$$

   where abs is the absolute value
3. *if G' is already k-degree anonymous, then $G' \neq G$*

The solution, G', is a graph representing a social network that

- is an anonymised network with k-degree anonymity property that
  - hides user sensitive information;
  - makes it more difficult to de-anonymise the network.
- is as close as possible to the original which implies that the information has fewer changes. Hence, any study performed on it will be significant.

5 Implementation of ADegree

We implemented ADegree in Gecode [5]. The implementation models the problem as a Constraint Satisfaction Problem (CSP). It finds an anonymised graph from a given social network. The found graph protects sensitive information whilst preserving the structure of the graph as much as possible. Gecode is a powerful tool for solving CSPs. The idea behind Constraint Programming is to define the model as a set of constraints over finite domain variables. Gecode framework will prune variables' domain until it finds a solution. A solution is found whenever the domain of all variables in the model has been reduced to a value. A user defines how to prune variables' values by specifying branching rules.

Part of previously described formal definitions are implemented in this study since our current implementation does not take into account labelled graphs. Subsection 5.2 is devoted to explain the possible implementations. Current implementation of ADegree can be found in [18].

5.1   ADegree

When working with Constraint Programming, one needs to define the followings.

- finite domain variables;
- constraints to be satisfied by the value of the variables;
- branching rules and
- the search engine.

Finite Domain Variables: ADegree is a data structure that represents a k-degree anonymous undirected-graph. ADegree is represented by a set of adjacent nodes.

Constraints: The set of constraints that needs to be satisfied is that the resulting graph is

cons1: a super-graph of the input graph and
cons2: k-degree anonymous,
cons3: as close as possible to the input graph.

In order to ensure that **cons1** holds it is necessary to add the following constraint.

- For every edge that is represented by pair of nodes (a, b), a is in an adjacent set of b and b is in an adjacent set of a.

In order to ensure that **cons2** holds we need to know that for every node there is at least $k - 1$ nodes with the same amount of adjacent nodes. We introduced a degree sequence variable and constraints are as following.

- The size of the degree sequence is the number of nodes in the graph.
- Each value in the degree sequence corresponds to the cardinality of the corresponding adjacent set.
- For every value of degree there is at least $k - 1$ degrees with the same value.

Branches: It is important to define the branch strategy to be able to find a solution in an acceptable time. For instance, choosing any branch strategy on the edges of the graph and running Depth First Search (DFS) engine will yield a solution. However, the search engine will try, for every possible super-graph, to check whether it has a k-degree anonymous sequence or not. This approach is feasible on small graphs but the time complexity will increase significantly as the graph increase its size.

In order to increase performance several methods can be applied. The first technique is to optimise the constraints. Gecode can cut off the search space by defining stronger conditions. One way to define such stronger constraints is to define a sorted degree sequence as realizable. That is, there exists a simple graph whose nodes have precisely the sorted degree sequence. Erdös and Gallai described the necessary and sufficient condition for a degree sequence to be realizable [4]. These conditions are n inequalities on sorted degree sequence. It can be achieved by using sorted versions of numerical arrays provided by Gecode.

For every $l \in [1, n-1]$

$$\sum_{i=1}^{l} d_i \leq l(l-1) + \sum_{i=l+1}^{n} min(l, d_i)$$

where d_i is i-th element in a sorted degree sequence. $min(l, d_i) \leq d_i$ we can transform it into weaker but faster condition:

$$\sum_{i=1}^{l} d_i \leq l(l-1) + \sum_{i=l+1}^{n} d_i$$

This solution has an implicit advantage. To find the existence of k same degree values in a sorted degree sequence, it is enough to search only in $k-1$ surrounding, not in the whole sequence. Such constraints are much lighter because they are from $2k-1$ variables and not from n.

All previous improvements deal only with degree sequence and with its sorted version but not with edges. Our implementation does not take into account pruning of edges. Introducing edges to the branching strategy will slow down the execution if the branch strategy is not carefully defined. Section 5.2 proposes how to solve the problem. The main problem is that the search engine checks several graphs with the same degree sequence even if they are not k-degree anonymous. Hence, the branch strategy needs to be defined so that it branches first on the degree sequence (most of the constraints make restrictions on a degree sequence) rather than on the edges. This will make the search run faster. Gecode allows users to define any order of search. The following is the ideal searching order:

1. Branch on sorted degree sequence
2. Branch on degree sequence
3. Branch on edges

Search: Gecode provides two search engines: Depth First Search (DFS) and Branch and Bound (BAB). DFS gives a solution which satisfies all the constrains but it cannot compare solutions to provide the best one. BAB is designed to give the best solution. For graph anonymisation, the idea is to find a graph that is as close as possible to the initial input. This gives a cost function that can be used to compare results given by the search engine. The cost function is defined by a user and corresponds to the Query of Interest (QI). Once the QI is defined, BAB gives some solution in increasing (decreasing) order until it achieves the best one.

Depending on the cost function, different strategies of searching within one structure can give different execution times. For example, if the cost function is the sum of the degree sequence and its goal is to minimize the sum (e.g. to take the least possible value of a degree first), the solution can be found faster.

5.2 Towards Edge Labelling

This section is devoted to express how to improve the implementation in order to consider labelling. The main idea is to transform this new problem into the previous one by defining the degree sequence as a tuple of degrees for every label. We need to be sure that every node appears either 0 or at least k times in the graph. If there are m labels, each degree is represented by $(d_1, d_2, \ldots, d_m)$ where d_i is the degree of i-th label of node. This tuple can be written as $deg = d_1 + d_2 K + d_3 K^2 + \ldots + d_m K^{m-1}$. Where K is a large integer that is bigger than maximum possible degree value. For example, if we assume that there can be only one edge with same label from node a to b, then K can be $n-1$ (where n is the length of the sequence). By doing this transformation we uniquely assign an integer number to any possible combination of a degree sequence.

Applying this transformation we can now anonymise undirected graphs with labelled edges. Now we can transform this variation to the directed graph case. The only thing needed is to assign different labels to the beginning of the edge and to its end.

6 Using ADegree

Consider the graph $G(V, E)$ depicted in Fig. 4. The graph shows an excerpt of a social network representing a AT&T network. Nodes represent users, arcs represent calls made between two users, and labels represent time of the calls.

A typical scenario in which AT&T might be interested is studying the graph to come up with better call plans targeting a group of users. The scenario also assumes that AT&T outsources this study (necessity of anonymised the graph). The query of interest (i.e. $QI = \langle G, P, T, F \rangle$) is to find out what is the average number of calls made by a type of user (e.g., teenagers) during a period of time (e.g., peak hours), in order to make a specific plan of minutes. Users are classified by their ages and the calls are classified by the period of time. Let P be the set of possible groups of ages and T be the set of possible periods of time as shown in Table 1.

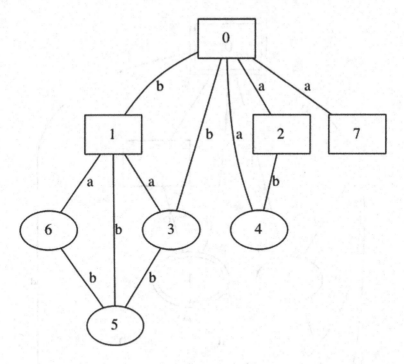

Fig. 4. Modified AT&T network.

Table 1. Classification of the AT&T social network. (a) Set P (b) Set T

(a)   Set P

Nodes	Type
$\{0,1,2,7\}$	p_1
$\{4,8\}$	p_2
$\{3,5\}$	p_3
$\{6\}$	p_4

(b)   Set T

Arcs	Type
$\{a\}$	t_1
$\{b\}$	t_2

Function F is formally defined as, $\forall p, t \cdot p \in P \wedge t \in T \Rightarrow$

$$F(G^-(p,t)) = \frac{\sum_{v_i \in V^-} \left| \{v_2 \mid v_2 \in V^- \wedge (v_1, v_2) \in E^-\} \right|}{|V^-|}$$

where $G^-(p,t) = (V^-, E^-)$ is the restricted graph obtained from G when restricting the set of nodes to those of classification p and the set of arcs to those associated with classification t. An example of the weight associated to the graph taking the type of user teenagers (p_1) during a period of time 5 min (t_1) is $F(G^-_{p_1,t_1}) = 0.5$.

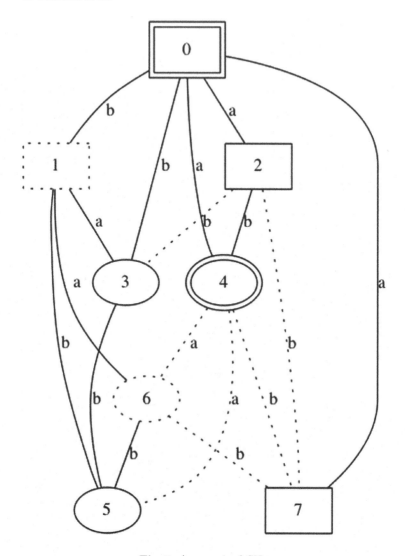

Fig. 5. Anonymised SN.

The anonymised Social Network found by `ADegree` is depicted in Fig. 5. The graph is a 2-degree anonymous. Dotted edges were added by our solution where shapes of nodes denote equivalence classes.

7   Conclusions

The use of social networks has increased exponentially and so as the information shared by owners of these networks to different entities. This information is of special interest for many companies to conduct studies such as marketing or epidemiology. The owners of the networks are compelled by law to protect the private information of their users when the network is sold/shared/released. Anonymisation methods play a key role in this cases. The main purpose of these methods is to protect the sensitive (private) information of the users involved in the network while making the process of de-anonymisation more difficult. Current methods for anonymisation are not safe enough. We have formally introduced the ADegree constraint utilized in the process. The constraint seeks to protect sensitive information of the network avoiding issues of the current anonymisation processes. ADegree finds a network G' such that: a) private information is removed from G', making the network anonymous. As stated before, this process is not enough to protect the network from different attacks. We use anonymisation technique (i.e., adding fake edges), enforcing k-degree anonymous property over G'. b) G' has some modification with respect to the original network in order to make the reverse process of de-anonymisation harder. Enforcing this property implies that G' losses some interesting properties that were contained in the original network. The process of finding G' takes into account the utility of the network, that is, G' is as close as possible to original graph, where closeness is defined by users (a parameter of the constraint). We also show the implementation of the constraint using Gecode. The current implementation does not consider labelled graphs. Section 5.2 proposes an extension of the current implementation to handle labels whilst not introducing extra time complexity. We plan to extend the implementation to handle labels to compare our approach to the existing ones. We also plan to formally prove the following properties of ADegree.

Correct: it never removes values that are consistent with respect to its constraint.

Checking: it is singleton correctness (accept all satisfying assignments) and singleton completeness (reject all non-satisfying assignments).

Domain consistent: it removes all inconsistent values.

Finally, we plan to exploit all CP techniques in order to anonymise a social network under different type of attacks such as k-automorphism, l-neighbours and others. Also, we started investigating to provide a solution for preserving privacy under sequential releases of the same Social Network.

References

1. Backstrom, L., Dwork, C., Kleinberg, J.: Wherefore art thou r3579x?: anonymized social networks, hidden patterns, and structural steganography. In: Proceedings of the 16th International Conference on World Wide Web, WWW 2007, pp. 181–190. ACM, New York (2007)
2. Bhagat, S., Cormode, G., Krishnamurthy, B., Srivastava, D.: Class-based graph anonymization for social network data. Proc. VLDB Endow. **2**(1), 766–777 (2009)
3. Campan, A., Truta, T.M.: Data and structural k-anonymity in social networks. In: Bonchi, F., Ferrari, E., Jiang, W., Malin, B. (eds.) PInKDD 2008. LNCS, vol. 5456, pp. 33–54. Springer, Heidelberg (2009). https://doi.org/10.1007/978-3-642-01718-6_4

4. Erdős, P., Gallai, T.: Graphs with prescribed degrees of vertices. Mat. Lapok **11**, 264–274 (1960). Hungarian

5. Gecode Team: Gecode: Generic constraint development environment (2006). http://www.gecode.org

6. Hay, M., Miklau, G., Jensen, D., Towsley, D., Li, C.: Resisting structural re-identification in anonymized social networks. VLDB J. **19**(6), 797–823 (2010)

7. Lebedev, A., Lee, J.Y., Rivera, V., Mazzara, M.: Link prediction using top-k shortest distances. In: Calì, A., Wood, P., Martin, N., Poulovassilis, A. (eds.) BICOD 2017. LNCS, vol. 10365, pp. 101–105. Springer, Cham (2017). https://doi.org/10.1007/978-3-319-60795-5_10

8. Lee, J.Y., Oh, J.C.: Agent perspective social networks: distributed second degree estimation. In: Alhajj, R., Rokne, J. (eds.) Encyclopedia of Social Network Analysis and Mining, pp. 1–12. Springer, New York (2017). https://doi.org/10.1007/978-1-4614-7163-9_110187-1

9. Lee, J.Y.: Reputation computation in social networks and its applications (2014)

10. Lee, J.Y., Lopatin, K., Hussain, R., Nawaz, W.: Evolution of friendship: a case study of mobiClique. In: Proceedings of the Computing Frontiers Conference, pp. 267–270. ACM (2017)

11. Lee, J.Y., Oh, J.C.: A node-centric reputation computation algorithm on online social networks. In: Kazienko, P., Chawla, N. (eds.) Applications of Social Media and Social Network Analysis. LNSN, pp. 1–22. Springer, Cham (2015). https://doi.org/10.1007/978-3-319-19003-7_1

12. Liu, C., Mittal, P.: Linkmirage: How to anonymize links in dynamic social systems. CoRR, abs/1501.01361 (2015)

13. Liu, K., Terzi, E.: Towards identity anonymization on graphs. In: Proceedings of the 2008 ACM SIGMOD International Conference on Management of Data, SIGMOD 2008, pp. 93–106. ACM, New York (2008)

14. Mazzara, M., Biselli, L., Greco, P.P., Dragoni, N., Marraffa, A., Qamar, N., de Nicola, S.: Social Networks and Collective Intelligence: A Return to the Agora. IGI Global, Hershey (2013)

15. Narayanan, A., Shmatikov, V.: De-anonymizing social networks. In: Proceedings of the 2009 30th IEEE Symposium on Security and Privacy, SP 2009, pp. 173–187. IEEE Computer Society, Washington (2009)

16. Nguyen, B.P., Ngo, H., Kim, J., Kim, J.: Publishing graph data with subgraph differential privacy. In: Kim, H., Choi, D. (eds.) WISA 2015. LNCS, vol. 9503, pp. 134–145. Springer, Cham (2016). https://doi.org/10.1007/978-3-319-31875-2_12

17. Rossi, F., van Beek, P., Walsh, T.: Handbook of Constraint Programming (Foundations of Artificial Intelligence). Elsevier Science Inc., New York (2006)

18. Solonets, S.: Adegree constraint implementation (2017). https://github.com/Solonets/ADegree

19. Sweeney, L.: K-anonymity: a model for protecting privacy. Int. J. Uncertain. Fuzziness Knowl.-Based Syst. **10**(5), 557–570 (2002)

20. Zhou, B., Pei, J.: Preserving privacy in social networks against neighborhood attacks. In: Proceedings of the 2008 IEEE 24th International Conference on Data Engineering, ICDE 2008, pp. 506–515. IEEE Computer Society, Washington (2008)

21. Zou, L., Chen, L., Tamer Özsu, M.: K-automorphism: a general framework for privacy preserving network publication. PVLDB **2**(1), 946–957 (2009)

JEREMIE: Joint Semantic Feature Learning via Multi-relational Matrix Completion

Jiaming Zhang[1], Shuhui Wang[1(✉)], Qiang Qu[2], and Qingming Huang[1]

[1] Key Lab of Intelligent Information Processing of Chinese Academy of Sciences (CAS), Institute of Computing Technology, CAS, Beijing 100190, China
jiaming.zhang@vipl.ict.ac.cn, wangshuhui@ict.ac.cn, qmhuang@ucas.ac.cn
[2] The Global Center for Big Mobile Intelligence, Frontier Science and Technology Research Centre, Shenzhen Institutes of Advanced Technology, CAS, Shenzhen 518055, China
qiang@siat.ac.cn

Abstract. The relations among heterogeneous data objects (e.g., image, tag, user and geographical point-of-interest (POI)) on interactive Online Social Media (OSM) play an important information source in describing complicated connections among Web entities (users and POIs) and items (images). Jointly predicting multiple relations instead of single relation completion in separate tasks facilitates sufficient knowledge sharing among heterogeneous relations and mitigate the information imbalance among different tasks. In this paper, we propose JEREMIE, a **J**oint S**E**mantic Featu**R**e L**E**arning model via **M**ulti-relational Matr**I**x Compl**E**tion, which jointly complements the semantic features of different entities from heterogeneous domains. Specifically, to perform appropriate information averaging, we first divide the social image collection into data blocks according to the affiliated user and POI information, where POIs are detected by mean shift from the GPS information. Then we develop a block-wise batch learning method which jointly learns the semantic features (e.g., *image-tag*, *POI-tag* and *user-tag* relations) by optimizing a transductive matrix completion framework with structure preservation and appropriate information averaging functionality. Experimental results on automatic image annotation, image-based user retrieval and image-based POI retrieval demonstrate that our approach achieves promising performance in various relation prediction tasks on six city-scale OSM datasets.

Keywords: Multi-relational learning · Semantic feature
Matrix completion · Information retrieval

1 Introduction

Multiple relations among heterogeneous data objects (e.g., image, tag, user and point-of-interest) on interactive Online Social Media (OSM) play an important role in describing complicated connections among entities (user and point-of-interest) in real world and items (image) on the Internet. In this context, diversified information acquisition needs have been evoked by the users for enjoying

© Springer International Publishing AG 2018
C. Doulkeridis et al. (Eds.): MATES 2017, LNCS 10731, pp. 87–108, 2018.
https://doi.org/10.1007/978-3-319-73521-4_6

better daily life. For example, users tend to find potential friends with similar preference on the image content and semantics. They may be pleased if the recommended points-of-interest (POIs) contain the locations actually meet their purposes. Therefore, it has become more and more emergent to develop effective and efficient methodologies for heterogeneous data object retrieval.

In this paper, we prescribe the heterogeneous data object retrieval problem into several relation prediction tasks, e.g., automatic image annotation (*image-tag*), image-based user retrieval (*image-user*), image-based POI retrieval (*image-POI*), etc. Traditionally, the statistical dependency among certain types of data objects are modeled by specific learning tasks. For example, to obtain a sufficient number of high-quality tags for social images based on existing manually labeled tags, tag refinement [12,14,30], tag completion [13,25,29] and multi-label learning [8] are widely studied to deal with the absence and noise in the collected social tags [6]. Recommendation models are employed to discover the relation between users and content [16,19,24]. Then for POI recommendation, current works [3,27,28] mainly focus on leveraging the geographical and social properties to improve recommendation effectiveness. The correlation among heterogeneous objects on OSM is uniformly modeled with multi-layered graphs [15]. All of these works either focus on a specific single relational learning task or use a unified graph structure to represent heterogeneous objects and their correlation.

In fact, entities like users and POIs or web items like images are all associated with tags with rich semantics. This common semantic attribute makes it possible to facilitate the knowledge sharing and propagation by jointly modeling the relation and interaction between heterogeneous data objects (image, user and POI) and semantic tags. Specifically, we regard the tag-oriented relation X-tag (X can be user, POI or image) as the semantic feature of each data object, which reflects the users' interest in the tagging activities for *user-tag* and POI properties for *POI-tag*, visual content for *image-tag* respectively. The semantic feature of a given entity can be regarded as a representation in the shared semantic space. In addition, each image has a visual feature to represent the visual content, and the content similarity is regarded as the core structure to determine how labels are propagated in standard image tagging models.

Besides the semantic feature, there exist rich attributes among OSM images. For example, each image is uploaded by a single user, while a user may have uploaded multiple images. Each image belongs to a detected POI, while a POI may have multiple images within its geographical range. The social context existing in the attributes of heterogeneous objects can be adopted to deploy the multi-relational learning tasks. For example, the image location may be strongly related to the geographical tags, and may also have direct implication on the scene and object in the image [5,16,18]. From the users' perspective, users with similar social attributes tend to co-occur in the same places, and they are more likely to be interested in the same type of images [25]. Since the rich attributes describe natural cluster structure of the OSM images, we divide the OSM image dataset into a set of relation data blocks according to the user structure and POI structure. Then the learning process of the X-tag relation can be performed by cluster-specific block matrix completion and factorization, which are actually information averaging mechanisms by latent factor modeling among the given relation sub-matrices within an appropriate POI-specific or user-specific range. Furthermore, by constructing block-based relation learning mechanism, the computational efficiency can be enhanced, especially the reduced storage consumption. Consequently, the multi-relational matrix completion problem on large scale OSM data can be solved with parallel optimization.

In our research context, the semantic feature of each entity is noisy and sparsely observed due to the complicated user behaviors [7], which brings great challenges in quantifying the true relations among heterogeneous data objects. Therefore, considering the intrinsic sharing nature among these relations, we jointly learn the semantic features from their corresponding noisy and sparse observation in a multi-relational matrix completion framework. Our goal is to learn representative semantic features for user, POI and image, regularized by visual similarity among images and the relation data block structure. The *image-user* and *image-POI* relation can be reconstructed based on the learned model by calculating the inner product on the corresponding learned semantic features, which is similar in spirit as the reconstruction procedure in classical matrix factorization approaches. Based on the reconstructed relations, the *image-user* and *image-POI* retrieval can be performed by relation score ranking.

Accordingly, our framework consists of two steps, one is the preprocessing step dealing with the data partition by users or POIs, and the second step is the semantic feature learning step where the semantic features are learned by the proposed multi-relational matrix completion model. In the preprocessing step, we individually separate images into data blocks with respect to users and POIs, respectively. Then the whole learning procedure can be formulated into a set of sub-tasks. In the learning step, the unified objective loss function contains two main modules, i.e., a structure preservation module using grouped affiliation relations and an information averaging module using visual feature of images. The structure preservation module penalizes the difference defined by the entity similarity calculated by the semantic feature between user and image with the assistance of the cluster structure. The information averaging module adopts the same formulation with the structure preservation module. However, the information averaging module employs both the visual feature and the semantic feature of image. Consequently, the relations among heterogeneous objects can be more reliably learned than other relational learning models [14, 21, 25]. Moreover, our method better reflects the intrinsic nature of heterogeneous object correlations than unified graph-based model [15], thus the relations can be learned more accurately.

The contribution of our framework is listed as below:

– We propose the JEREMIE approach, a novel Joint sEmantic featuRe lEarning fraMework via Multi-relational MatrIx complEtion which involves both mapping relationship for structure preservation and intrinsic feature for information averaging.

- The data partition step builds a localized neighborhood environment for more compact semantic feature learning, which is much closer to the intrinsic data distribution for information averaging within local neighbors compared with other irrelevant samples.
- The better completion results of semantic features of images promote that the learned semantic features of users and POIs to have better representation for their properties in return, which indicates better performance of the reconstruction of the *image-user* and *image-POI* relation.
- The relation structure in our proposed framework makes better use of rich social context among multi-relational domains than unified graph-based models and collective learning algorithms.

In summary, by combining nonnegative matrix factorization and transductive matrix completion approaches, we propose JEREMIE to jointly complements the semantic features of different entities from heterogeneous domains. An efficient alternating sub-gradient descent algorithm can be adopted to simultaneously learn the POI-specific and user-specific relation matrix blocks in an alternating optimization style. Experiments on real-world OSM datasets demonstrate the promising power of our approach. The framework is illustrated in Fig. 1.

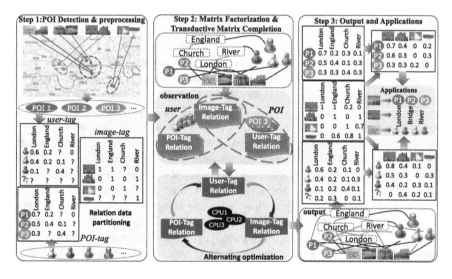

Fig. 1. Framework of our proposed method JEREMIE (best viewed in color). JEREMIE consists of two steps. The first step is the data partition preprocessing step dealing with GPS locations (by Mean Shift procedure) and users of images. The second step is the proposed block-wise multi-relational matrix completion model for joint semantic feature learning. (Color figure online)

2   Related Work

In this section, we first briefly survey the existing approaches from the perspective of both the application (automatic image annotation, attribute-based social media analysis) and the multi-relational learning theory, and motivate our study for designing multi-relational matrix completion method.

2.1   Automatic Image Annotation

Besides general content-based automatic image annotation techniques [20], there are also many recent works exploiting multi-label learning techniques [8] to deal with image annotation problem.

Several researchers solve the image annotation problem by image retagging, tag recommendation, tag propagation, etc. Li et al. [12] propose a neighbor voting method for social tagging. Guillaumin et al. [6] propose a tag propagation (TagProp) method to transfer tags through a weighted nearest neighbor graph.

All the approaches mentioned above usually need high-quality annotated tags and good class assignments for confident model training, contrary to the reality that manually annotated tags contain many incorrect and noisy ones. Furthermore, matrix completion techniques are also introduced to address poor initial image annotation problem [13,25], getting convincible experiment results. More recently, Wu et al. [25] build a concise tag matrix completion computational framework based on both semantic-content and tag co-occurrence consistency. Motivated by this work, we adopt matrix completion in our proposed joint semantic feature learning framework.

2.2   Attribute-Based Social Media Analysis

In social media analysis, many algorithms for social user recommendation have been proposed in recent years. Combining users and visual content, social recommendation models are studied to discover the connection between users and content [19,24]. From the perspective of correlation among heterogeneous objects in OSM, graph-based approaches are utilized to uniformly model the recommendation problems [15]. These correlation based recommendation algorithms usually build on global relationship, which will inevitably encounter computational complexity problems as the scalability goes up.

Besides numerous research works on social media user analysis, there are many research works focusing on combining geographical attributes and visual content. To analyze large scale online image collections with both geographical and visual content information, Crandall et al. [5] formularize the image location estimation as a classification problem by classifying images into POI categories. Moxley et al. [18] adopt geographical based search strategy to provide candidate tags and images which are similar in visual content. Liu et al. [16] propose a unified framework using subspace learning in personalized and geo-specific tag recommendation for images on OSM.

2.3   Multi-relational Learning

User-item clustering or matrix factorization techniques in collaborative filtering have been successfully utilized to represent non-trivial similarities between the relation patterns of entities in single relational data. Easily extended to multi-relational learning, several clustering based approaches have been proposed. Sutskever et al. [22] use a non-parametric Bayesian clustering of entities embedding in a collective matrix factorization formulation. Zhu et al. [31] refine the clustering setting to allow entities to have a mixed clusters membership. To share parameters between relations, [22,31] build models that cluster not only entities but relations as well.

Similarly, as pointed out by [9], a lot of methods for multi-relational data have been designed within the framework of relational learning from latent attributes, which learn latent embeddings of the entities by operating on relations through factorization (matrix or tensor) framework. The fact that relations can be similar or highly related suggests that a superposition of independently learned models for each relation would be highly inefficient especially since the relationships observed for each relation are extremely sparse. The basic idea of collectively learning several relations simultaneously is to share the common embedding via collective matrix factorization (CMF) [21]. The insight we model pairwise relations is similar to CMF [21]. However we adopt transductive matrix completion instead of CMF in a similarity propagation manner, which is quite different from CMF.

3   Notations and Preliminaries

We denote n as the number of images uploaded by u users, m as the number of unique tags in the dataset and p as the number of clustered POIs. Then the two types of relation matrices are denoted as:

Observed Image Semantic Feature: $\widehat{T} \in \{0,1\}^{n \times m}$ is a binary matrix, where $\widehat{T}_{st}$ is set to 1 if tag t is assigned to image s and otherwise 0. $\widehat{T}$ represents the tagging behavior on each image, while the tags of an image provide some weak annotations on the diversified visual content.

Observed User Semantic Feature: $\widehat{U} \in \mathbb{R}^{u \times m}$ is the observed user-tag matrix, where $\widehat{U}_{kt} = \frac{1}{|k_n|} \sum_s \widehat{T}_{st}$ and $|k_n|$ denotes the number of user k's images. It can be considered as a normalized histogram of tags for users.

Observed POI Semantic Feature: $\widehat{P} \in \mathbb{R}^{p \times m}$ is the observed POI-tag matrix, where $\widehat{P}_{lt} = \frac{1}{|l_n|} \sum_s \widehat{T}_{st}$ and $|l_n|$ denotes the number of images in the l-th POI. Similarly, it can be considered as a normalized histogram of tags for POIs.

Image-User Relation Matrix: $R \in \{0,1\}^{u \times n}$ is the binary user-image matrix that represents the affiliation between users and images, where R_{ks} is 1 if image s belongs to user k and otherwise 0.

Image-POI Relation Matrix: $S \in \{0,1\}^{p \times n}$ is the binary POI-image matrix that represents the aggregation distribution between POIs and images, where S_{ls} is 1 if image s belongs to POI l and otherwise 0.

The goal of our proposed framework is to obtain $\widetilde{R}$, $\widetilde{S}$, T, U and P by completing the three corresponding original observed matrices $\widehat{T}$, $\widehat{U}$ and $\widehat{P}$, where $\widetilde{R} = UT^\top$ and $\widetilde{S} = PT^\top$. Each entry T_{st} in completed image-tag matrix T indicates the probability of assigning tag t to image s, with U_{kt} in U the probability of attaching tag t to user k and P_{lt} in P the probability of attaching tag t to POI l respectively.

For the representation of the visual content of images, we utilize the feature (the output of the fifth layer) of the deep convolutional neural networks (CNN) [10] trained on ILSVRC 2012 dataset. We denote the visual feature matrix as $V \in \mathbb{R}^{n \times d}$ where the i-th row corresponds to a d-dimension visual feature of image i.

4 Approach

First we introduce the data partition including the POI detection procedure as the preprocessing step of our algorithm. Then we provide details about our multi-relational matrix completion framework.

4.1 POI Clustering via Mean Shift

The points-of-interest (POIs) that we refer to in this paper is different from the definition based on users' check-in actions [28]. As discussed in previous section, geographically adjacent images may have similar visual content or semantic information with higher probability, which indicates potential POIs. Thus we apply a density-based clustering method Mean Shift [4] to detect POIs from the GPS information (latitudes and longitudes) of social images. With the POI detection results, each image obtains a new aggregated attribute POI. Then we can partition T into a set of sub-matrices according to POIs. After that we can build index for images based on the *image-POI* relation.

4.2 Data Partition with Users and POIs

Without loss of generality, we refer to the index of user as i and POI as j respectively for the interpretation of the data partition procedure. Within the two affiliation relations, *image-User* and clustered *image-POI*, we individually separate images into user blocks and POI blocks respectively. That means, the images which belong to user i (POI j) will be assigned to data block i (j). Then the relation matrices T, U and R will be partitioned into corresponding sub-matrices T_i, U_i and $R_i (i \in \{1,2,\cdots,u\})$ within user blocks, with T_j, P_j and $S_j (j \in \{1,2,\cdots,p\})$ for T, P and S within POI blocks respectively.

Algorithm 1. Optimization procedure of **JEREMIE**

Input: Observed Relation Matrices:$R_i, \widehat{T}_i, \widehat{U}_i, S_j, \widehat{T}_j, \widehat{P}_j$
Visual Feature Matrices: $V_i, V_j, i \in \{1, \ldots, u\}, j \in \{1, \ldots, p\}$
Output: Completed Relation Matrices: $\widetilde{R}, \widetilde{S}, T^t, U^t, P^t$;
Initial: $W_U^1 = W_P^1 = I_{d,m}, T_i^1 = \widehat{T}_i^1, U_i^1 = \widehat{U}_i^1, T_j^1 = \widehat{T}_j^1, P_j^1 = \widehat{P}_j^1, t = 1$;
while $\|\mathcal{L}^{t+1} - \mathcal{L}^t\| \geq \epsilon \mathcal{L}^t$ **do**
    Step size $\delta_t = \delta_0/t$;//**The loops below are both executed in block-wise**
    manner
    for $i = 1$ to u **do**
        Calculate $\overline{T}_i^{t+1}, \overline{U}_i^{t+1}, \overline{W_U}^{t+1}$: Eq. 14
        Update and Concatenate T_i^{t+1}, U_i^{t+1}: Eq. 16
    end for
    for $j = 1$ to p **do**
        Calculate $\overline{T}_j^{t+1}, \overline{P}_j^{t+1}, \overline{W_P}^{t+1}$: Eq. 14
        Update and Concatenate T_j^{t+1}, P_j^{t+1}: Eq. 16
    end for
    Update W_U^{t+1} and W_P^{t+1}: Eq. 16
    t = t + 1;
end while
$\widetilde{R} = U^t T^{t\top}, \widetilde{S} = P^t T^{t\top}, T^t = \min\left(T^{Pt}, T^{Ut}\right)$
return $\widetilde{R}, \widetilde{S}, T^t, U^t, P^t$

4.3   JEREMIE Algorithm

Without loss of generality, we take user as the example to describe the details of our framework, and directly give the symmetric formulation for POI respectively.

Basic NMF Framework. Firstly proposed in [11], NMF is a commonly used matrix factorization technique which require the elements both in original matrix and factor matrices must be nonnegative. In our framework, we decompose the relation matrix R_i into two nonnegative latent factor matrices U_i and T_i aiming to find the semantics supervised reconstruction of R_i from tag space, represented as $R_i \approx U_i V_i^\top$. Usually, the approximation is quantified by a cost function constructed by the Frobenius norm denoted as $\|E_i\|_F^2$, where:

$$E_i = R_i - U_i T_i^\top. \tag{1}$$

Similarly, we obtain the formulation for POI denoted as $\|E_j\|_F^2$, where:

$$E_j = S_j - P_j T_j^\top. \tag{2}$$

In our framework, we force the $k = m$ to ensure the latent factor matrices are isomorphic with the tag matrices.

Modeling Tagging Preference via Structure Preservation. For all of the images uploaded by the same user in the same POI, they tend to assign the same tagging results from themselves and other users. The k-th row of U_i depicts the actual tag distribution of user k. Similarly, the k-th row in the product $R_i T_i$ reveals the weighted sum of image tag distribution for the images related to user k. So $R_i T_i$ can be regarded as a refined estimation for U_i using the group of images for each user. To formulate the tagging preference via user interest, we calculate the difference between the user-wise similarity $U_i U_i^\top$ and its estimation $R_i T_i T_i^\top R_i^\top$ with Frobenius norm. Then we define the structure preservation constraint term as $\|F_i\|_F^2$, where:

$$F_i = R_i T_i T_i^\top R_i^\top - U_i U_i^\top. \tag{3}$$

Similarly, we have the structure preservation constraint term for POI denoted as $\|F_j\|_F^2$, where:

$$F_j = S_j T_j T_j^\top S_j^\top - P_j P_j^\top. \tag{4}$$

Modeling Tagging Preference via Information Averaging. The visual content and annotated tags usually have intrinsic semantic relation. It makes great improvement to utilize visual information in image-tag matrix completion [13,25]. To enhance the semantic coherence between visual content and tags, we penalize the difference of similarities in visual feature space and textual semantic space with a Frobenius norm denoted as $\|T_i T_i^\top - V_i V_i^\top\|_F^2$.

However, low level visual features are less capable than tags for semantic representation of a given image. To reduce the semantic gap, we introduce a feature mapping matrix $W_U \in \mathbb{R}^{d \times m}$, which can directly map the visual feature into textual semantic space. Then the visual constraint term can be rewritten as $\|G\|_F^2$, where:

$$G_i = T_i T_i^\top - V_i W_U W_U^\top V_i^\top. \tag{5}$$

Similarly, we obtain the visual constraint term for POI denoted as $\|G_j\|_F^2$, where:

$$G_j = T_j T_j^\top - V_j W_P W_P^\top V_j^\top. \tag{6}$$

Regularization Terms. To avoid dense solution of semantic feature matrices, we require that only a small number of entries of T_i, T_j, U_i and P_j are nonzero, *i.e.*, a small number of unique tags are attached to each object (image, user or POI). As in many sparse coding literatures [17], we introduce two ℓ_1-norm regularization terms $\|T_i\|_1 + \|U_i\|_1$ to pursue a sparse solution of T_i and U_i, with $\|T_j\|_1 + \|P_j\|_1$ for POI respectively. For the shared mapping matrices W_U and W_P, we also add ℓ_1-norm regularization terms to enforce the sparsity.

Unified Loss Function. Finally, with respect to all of these criteria, we formulate our optimization problem as follows:

$$\underset{T_i, U_i, W_U, T_j, P_j, W_P}{\arg\min} \sum_{i=1}^{u} \mathcal{L}_i + \sum_{j=1}^{p} \mathcal{L}_j + \tau \left(\|W_U\|_1 + \|W_P\|_1 \right) \tag{7}$$

$$\mathcal{L}_i = \|E_i\|_F^2 + \alpha\|F_i\|_F^2 + \beta\|G_i\|_F^2 + \theta(\|T_i\|_1 + \|U_i\|_1) \tag{8}$$

$$\mathcal{L}_j = \|E_j\|_F^2 + \alpha\|F_j\|_F^2 + \beta\|G_j\|_F^2 + \theta(\|T_j\|_1 + \|P_j\|_1) \tag{9}$$

where $\alpha, \beta, \theta, \tau > 0$ are parameters whose values can be easily tuned in the cross-validation procedure.

4.4 Optimization

Without loss of generality, we take T_i and U_i within user i and W_U as an example to describe the details of our algorithm. According to symmetry, the same optimization strategy are applied to T_j and P_j within POI j and W_P respectively. As we can see, the objective function is non-convex because of several non-quadric regularization terms. Accordingly, we design an alternating optimization procedure using subgradient descent method to solve the problem on large scale OSM datasets.

On the other hand, we may get dense intermediate solutions $T_i^t, U_i^t, i \in \{1, \ldots, u\}$ if we directly use the alternating subgradient descent approach to solve the original problem. It will significantly increase the computational time cost in each iteration. To avoid this ill-posed condition, we decompose the objective function into two parts according to the composite function optimization method [1]. In particular, we construct an auxiliary function as below:

$$A_i = \|E_i\|_F^2 + \alpha\|F_i\|_F^2 + \beta\|G_i\|_F^2. \tag{10}$$

The subgradients of the auxiliary function with respect to each relation submatrix are:

$$\nabla_{T_i} A_i = 2E_i^\top U_i + 2\alpha R_i^\top F_i R_i T_i + 2\beta G_i T_i \tag{11}$$

$$\nabla_{U_i} A_i = 2E_i T_i + 2\alpha F_i U_i. \tag{12}$$

And the subgradients with respect to W are:

$$\nabla_{W_U} A_i = -2\beta(\sum_{i=1}^{u} V_i^\top G_i V_i)W_U. \tag{13}$$

Then we can compute the intermediate optimal solutions of each objective matrix from iteration t to $t+1$ referring to the auxiliary function by ($Q \in \{T_i, U_i, W_U\}$):

$$\overline{Q}^{t+1} = Q^t - \delta_t \nabla_{Q^t} A_i, \tag{14}$$

where δ_t is the step size. Based on these intermediate solutions and the rest of nonconvex regularization terms, we formulate the auxiliary optimization problems as follows ($Q \in \{T_i, U_i\}$ and $\varphi = \theta$, or $Q = W_U$ and $\varphi = \tau$):

$$Q^{t+1} = \arg\min_Q \frac{1}{2}\|Q - \overline{Q}^{t+1}\|_F^2 + \varphi\delta_t\|Q\|_1 \tag{15}$$

Combined with the intermediate solutions, we obtain the new solutions for auxiliary problems as follows:

$$Q^{t+1} = \max(\mathbf{0}, \overline{Q}^{t+1} - \varphi\delta_t). \tag{16}$$

Each variable in $\{T_i, U_i, W_U\}$ is alternatively updated during the t-th iteration.

Block-Wise Batch Processing. Here we discuss the whole optimization procedure in all of the user blocks, with the same strategy applying to POIs respectively. For T_i and U_i of user i, its calculation progress is independent from the corresponding matrices of other users. In order to achieve correct calculation results and reduce the complexity of our model, we make W_U shared by all of the sub-matrices V_i in parallel computation, with W_P for V_j respectively. Algorithm 1 illustrates the main steps in our solution for the optimization problem.

5   Experimental Evaluation

We evaluate the performance of our model **JEREMIE** on three object retrieval tasks: automatic image annotation (AIA), image-based user retrieval (IBUR) and image-base POI retrieval (IBPR).

5.1   Dataset and Experiment Settings

We use two Flickr datasets which are named after several international metropolises attracting crowds of tourists to conduct the experiments for AIA, IBUR and IBPR. All of these datasets are built by giving a geographical filtration to a large scale social image dataset YFCC100M[1][23] published by Yahoo Web Lab[2].

As studied in [26], the tag distribution among images is extremely imbalanced and the majority of tags belong to a few images. Then we rank the tags according to its number of annotated images and select top 1000 to serve as the vocabulary in experiment. Table 1 shows some statistics of these datasets.

Users' tagging behavior in specific POI may become uncertain as the geographical scope goes larger. So we fix the bandwidth parameter in Mean Shift as 0.005 empirically to achieve '*landmark-scale*' POI detection, as defined in [5]. This bandwidth parameter is in correspondence with 500 meters as the maximum geographical radius of the POIs detected in our experiments.

Table 1. Statistics of the datasets used in the experiments

Dataset	Image		User		Tag				
	#Total	#GPS	#User	#Image per user	#Tag	#Tag per image		#Image per tag	
						Mean	Max	Mean	Max
London	1,338,388	771,099	16,225	47.52	1,000	5.2	63	4,016.4	481,957
New York	1,210,094	732,555	15,344	47.74	1,000	5.5	71	4,031.5	299,867
Paris	532,562	425,282	10,510	40.46	1,000	4.7	45	2,012.5	342,987
Beijing	146,541	79,140	2,016	39.26	1,000	4.7	38	370.6	57,232
Shanghai	123,436	68,964	1,701	40.54	1,000	4.7	51	322.6	49,147

---

[1] https://multimediacommons.wordpress.com/.
[2] http://webscope.sandbox.yahoo.com/.

We utilize identity matrix to initialize the feature mapping matrices W_U and W_P. Then we randomly divide each dataset into training set with 90% images and testing set with the rest 10%, respectively. As we obtain T, $\widetilde{R}$ and $\widetilde{S}$ simultaneously, we share the same training set and testing set for all of the three tasks. We repeat the experiment five times and report the average performance, while each run adopts a new separation of the datasets. After the cross-validation procedure, we determine that $\alpha = 1000, \beta = 10, \theta = 1$ and $\tau = 1$. For TMC and other methods, we use the optimal parameter settings reported in related literature. The initial step size δ_0 is set as 10^{-7} empirically.

5.2 Compared Methods

Our proposed framework is evaluated on the three tasks with the following relevant algorithms:

- **FastTag (Fast)**: Baseline tagging method [2].
- **TagProp (Prop)**: Label propagation on weighted nearest neighbor graph [6].
- **TagRel (Rel)**: Tagging based on neighbor voting [12].
- **TMC**: Tag completion using content-semantic consistency and tag co-occurrence consistency [25].
- **PTC**: Tagging among POIs with users' tagging behavior prior [29].
- **CMF**: Multi-relational Learning method using collective matrix factorization [21].
- **JEREMIE**: our method.
- **JEREMIE_U (J-U)**: our method **only with user-related** constraint terms (with $\mathcal{L}_j$ removed in Eq. 7).
- **JEREMIE_P (J-P)**: our method **only with POI-related** constraint terms (with $\mathcal{L}_i$ removed in Eq. 7).
- **JEREMIE_F (J-F)**: our method **without visual content related** constraint terms (corresponding to $\beta = 0$ in Eqs. 8 and 9).
- **JEREMIE_G (J-G)**: our method **without structure related** constraint terms (corresponding to $\alpha = 0$ in Eqs. 8 and 9).

5.3 Automatic Image Annotation

The task of automatic image annotation is to automatically associate unlabelled images with semantically related tags. Given a query image q, we simply rank all the tags in descending order of their probability scores attached to image q, corresponding to the q-th row in T. Since we obtain T^U from user and T^P from POI simultaneously, we adopt the minimum pooling strategy

$$T_{st} = \min \left(T_{st}^P, T_{st}^U \right), \tag{17}$$

for the final tagging results to avoid very small positive completed entries (e.g. 1×10^{-8}). In particular, to test the robustness of our proposed method to the number of initial tags allocated to each image, we vary the number of initial training tags (denoted as e) for each image from $\{1, 2, 3, 4, 5\}$. Without loss of generality, we suppose image i has m_i manually annotated tags, corresponding to m_i non-zero entries in the i-th row of $\widehat{T}$ in training set. If $e \leq m_i$, we randomly select e tags as the initial annotation for image i. Otherwise if $e > m_i$, we remove image i from the training set. We use the Precision at TOP 20 (denoted as Precision@20) and Recall at TOP 20 (denoted as Recall@20) to measure the performance of different algorithms.

As shown in Tables 2 and 3, the annotation accuracy goes up with the increase of the number of initially observed tags in general for all methods. It can be explained by the fact that more observed tags for each image provides richer semantic information for knowledge propagation.

We also observe that all of our proposed methods outperform other compared methods. It provides evidence for that the matrix partition strategy makes both geographical locality and user interest consistency more compact in tag semantic space. Compared with state-of-the-art tagging methods (FastTag, TMC and TagProp), our algorithm utilizes auxiliary information such as the user-image and POI-image relations. The experiment results are dominated by the local consistency of tag distribution after matrix partition instead of the coherence between visual content and semantics.

5.4   Image-Based User and POI Retrieval

Closely related to friend recommendation, IBUR is a typical application scenario when a user of OSM wants to search possible users who has taken the query image, with IBPR searching for potential POIs respectively. Given a query image q, we simply rank all the known users in descending order of their relationship intensity scores to image q, corresponding to the q-th column in R and S for POI respectively. For each query image in the testing set, we expect the position of its original user or POI in the ranking list to be as high as possible. Basic settings such as initial training tags are the same with AIA, as our framework simultaneously outputs three X-tag relation matrices. In particular, we take Mean Reciprocal Rank (MRR) as performance measurement for different methods, which is defined as:

$$MRR = \frac{1}{|Q|} \sum_{q=1}^{|Q|} \frac{1}{Rank_q}, \qquad (18)$$

where Q is the testing set and $Rank_q$ is the position of original user or POI of testing image q in the ranked q-th column of R or S.

Table 4 shows the MRR scores of different methods on IBUR and IBPR. Since the two tasks are relatively independent of each other within the unified framework **JEREMIE**, we select the experiment results from corresponding branch, i.e. **J-U** for IBUR and **J-P** for IBPR respectively. We can see that the MRR scores of both **J-U** and **J-G** are even times larger than the baseline Matrix Factorization methods. While compared with general multi-relational learning method CMF, our algorithm adopts similarity consistency instead of

Table 2. Precision@20 of image tag recommendation

London	Precision@20						
Method	FastTag	TagRel	TagProp	CMF	TMC	PTC	JEREMIE
e = 1	16.67	16.26	16.92	19.43	26.16	35.19	**40.40**
e = 2	17.52	16.54	17.05	19.97	27.28	37.12	**42.26**
e = 3	17.37	17.07	17.21	19.29	27.63	37.39	**43.12**
e = 4	17.79	17.46	18.32	19.83	28.34	38.24	**43.50**
e = 5	18.61	17.73	18.62	20.19	28.05	38.38	**43.78**
New York	Precision@20						
Method	FastTag	TagRel	TagProp	CMF	TMC	PTC	JEREMIE
e = 1	13.13	12.67	15.51	16.77	17.93	23.79	**37.23**
e = 2	13.43	12.98	15.53	16.27	18.36	25.32	**38.14**
e = 3	13.88	13.15	15.90	16.83	18.38	26.14	**39.41**
e = 4	14.26	13.23	15.77	16.61	18.37	26.94	**39.87**
e = 5	14.68	13.62	15.96	17.02	18.40	27.06	**40.43**
Paris	Precision@20						
Method	FastTag	TagRel	TagProp	CMF	TMC	PTC	JEREMIE
e = 1	13.26	13.81	14.74	16.79	18.07	24.63	**33.68**
e = 2	13.63	12.90	14.61	16.12	18.72	25.56	**34.95**
e = 3	13.81	13.19	14.82	16.33	18.72	25.90	**35.51**
e = 4	14.36	13.57	15.05	16.64	18.89	25.90	**35.67**
e = 5	14.48	14.23	15.07	16.98	18.85	26.23	**35.91**
Beijing	Precision@20						
Method	FastTag	TagRel	TagProp	CMF	TMC	PTC	JEREMIE
e = 1	11.67	12.61	12.16	15.43	15.79	27.23	**33.34**
e = 2	12.15	13.21	13.74	15.92	15.70	29.23	**34.15**
e = 3	13.52	13.25	14.38	15.29	15.80	29.97	**34.96**
e = 4	13.40	13.56	14.42	15.77	15.67	30.01	**35.48**
e = 5	13.62	13.75	14.56	15.15	15.68	30.04	**36.72**
Shanghai	Precision@20						
Method	FastTag	TagRel	TagProp	CMF	TMC	PTC	JEREMIE
e = 1	11.84	12.01	11.70	15.29	15.94	23.07	**32.18**
e = 2	11.93	11.30	13.32	15.06	15.76	24.23	**34.13**
e = 3	12.66	12.46	13.59	15.51	15.83	24.90	**34.21**
e = 4	13.26	13.20	14.05	15.73	15.79	25.20	**35.78**
e = 5	14.20	12.86	14.20	15.33	15.78	25.32	**37.57**

Table 3. Recall@20 of Image Tag Recommendation

London	Recall@20						
Method	FastTag	TagRel	TagProp	CMF	TMC	PTC	JEREMIE
e = 1	25.16	23.27	27.50	33.76	36.28	63.69	**74.21**
e = 2	26.67	25.11	27.68	33.25	38.10	66.90	**78.26**
e = 3	26.88	26.37	28.38	33.98	38.63	67.98	**80.23**
e = 4	27.58	27.16	29.04	34.70	39.12	68.95	**81.08**
e = 5	28.43	27.86	29.15	34.33	39.97	69.28	**81.81**
New York	Recall@20						
Method	FastTag	TagRel	TagProp	CMF	TMC	PTC	JEREMIE
e = 1	28.66	26.32	33.64	37.91	42.15	54.17	**75.16**
e = 2	29.25	27.54	33.99	37.87	43.07	57.71	**79.29**
e = 3	29.34	28.41	34.57	38.02	43.18	59.72	**81.37**
e = 4	29.96	28.80	34.36	37.45	43.20	61.37	**82.56**
e = 5	30.84	29.66	34.91	37.99	43.17	61.67	**83.33**
Beijing	Recall@20						
Method	FastTag	TagRel	TagProp	CMF	TMC	PTC	JEREMIE
e = 1	32.97	33.77	33.62	43.26	43.35	71.91	**85.85**
e = 2	33.98	34.03	37.90	43.90	42.91	76.65	**89.80**
e = 3	36.88	34.36	39.54	43.53	43.18	78.24	**90.98**
e = 4	36.90	33.99	39.64	43.83	42.81	78.55	**91.26**
e = 5	39.97	33.86	40.03	43.01	42.82	78.51	**92.96**
Paris	Recall@20						
Method	FastTag	TagRel	TagProp	CMF	TMC	PTC	JEREMIE
e = 1	38.91	39.81	42.23	46.24	53.81	67.89	**84.74**
e = 2	39.56	38.16	42.14	45.95	54.91	70.15	**87.96**
e = 3	40.15	38.19	42.36	45.44	47.19	70.71	**89.24**
e = 4	41.47	39.57	43.84	45.77	48.29	71.05	**89.75**
e = 5	42.18	40.23	43.97	45.38	48.32	71.52	**90.24**
Shanghai	Recall@20						
Method	FastTag	TagRel	TagProp	CMF	TMC	PTC	JEREMIE
e = 1	32.09	31.50	32.64	41.64	43.98	64.36	**81.84**
e = 2	33.38	30.56	36.49	41.17	43.55	66.46	**84.87**
e = 3	35.80	32.66	37.06	42.07	43.76	68.68	**85.82**
e = 4	37.45	32.77	38.30	41.55	43.62	69.00	**86.57**
e = 5	38.39	33.54	38.76	41.93	43.62	69.11	**87.82**

Table 4. MRR performance of IBUR and IBPR

London	MRR @ IBUR				MRR @ IBPR			
Method	NMF	CMF	J-G	J-U	NMF	CMF	J-G	J-P
$e=1$		12.41	19.70	**33.46**		6.67	13.16	**20.87**
$e=2$		12.15	20.88	**34.70**		6.38	15.40	**22.74**
$e=3$	7.96	12.69	21.37	**35.63**	3.67	6.89	15.88	**23.34**
$e=4$		13.01	21.96	**36.93**		7.27	16.66	**23.86**
$e=5$		13.27	23.04	**38.68**		7.34	17.83	**24.95**
New York	MRR @ IBUR				MRR @ IBPR			
Method	NMF	CMF	J-G	J-U	NMF	CMF	J-G	J-P
$e=1$		16.44	26.13	**41.83**		10.62	17.17	**23.76**
$e=2$		16.62	28.36	**43.10**		10.34	18.45	**25.55**
$e=3$	11.07	16.86	28.78	**43.72**	6.41	10.85	18.90	**26.31**
$e=4$		17.11	29.66	**44.26**		11.20	19.33	**27.15**
$e=5$		17.48	31.05	**46.65**		11.36	20.62	**28.88**
Paris	MRR @ IBUR				MRR @ IBPR			
Method	NMF	CMF	J-G	J-U	NMF	CMF	J-G	J-P
$e=1$		24.11	33.52	**51.65**		11.43	17.62	**28.61**
$e=2$		23.98	34.49	**52.59**		11.05	18.27	**28.91**
$e=3$	15.34	24.45	34.52	**54.01**	5.56	11.77	19.03	**30.89**
$e=4$		24.76	35.10	**56.30**		12.06	20.50	**33.08**
$e=5$		25.22	36.34	**57.61**		12.31	21.41	**35.24**
Beijing	MRR @ IBUR				MRR @ IBPR			
Method	NMF	CMF	J-G	J-U	NMF	CMF	J-G	J-P
$e=1$		11.12	36.52	**46.32**		27.32	38.24	**56.34**
$e=2$		10.99	36.15	**47.21**		26.64	39.95	**56.32**
$e=3$	7.58	11.46	37.49	**48.37**	18.25	27.39	41.01	**58.81**
$e=4$		11.37	38.37	**48.70**		26.91	41.56	**59.63**
$e=5$		12.03	38.23	**48.85**		27.88	41.77	**59.92**
Shanghai	MRR @ IBUR				MRR @ IBPR			
Method	NMF	CMF	J-G	J-U	NMF	CMF	J-G	J-P
$e=1$		14.05	18.02	**38.28**		18.37	36.89	**40.48**
$e=2$		13.65	19.43	**41.04**		18.62	37.45	**42.83**
$e=3$	10.01	14.21	19.69	**41.67**	12.13	17.80	38.76	**44.86**
$e=4$		14.16	20.80	**43.87**		18.45	40.87	**47.74**
$e=5$		14.64	20.31	**43.81**		18.92	40.12	**47.27**

shared latent embedding. The performance gap between **J-G** and **J-U** in the IBUR task gives a strong support for the effectiveness of structure preservation in the procedure of jointly learning semantic features. Similar conclusion can be derive with **J-G** and **J-P** in the IBPR task. This experimental phenomenon demonstrates that the semantic features T_i and U_i are indispensable for the reconstruction of R_i in IBUR task, and we can derive similar conclusion in IBPR. Moreover, both the visual contents and structure information make contributions to the high-level semantic information sharing with users and POIs.

5.5 Evaluation of Model Effectiveness

To evaluate the effectiveness of the learning model of **JEREMIE**, we propose to consider different combinations of components by setting the corresponding parameters α and β. Since our learning model is symmetric, we first generate two individual sub-models **JEREMIE$_U$** and **JEREMIE$_P$** from the original model by wiping out corresponding components. Then we consider to test the effectiveness of structure preservation in **JEREMIE$_G$** by setting the parameter α to 0. Similarly, we obtain **JEREMIE$_F$** to test the effectiveness of information averaging via visual content by setting the parameter β to 0. All of the four types of weakening variant of our method have been introduced in Sect. 5.2.

Table 5 shows the scores of all tasks for this experiment in *Beijing* and *Shanghai*. In AIA task, we can see that the performance of fusion results with our **JEREMIE** model are better than separate results (**J-U** and **J-P**). It gives strong support for the effectiveness of minimum pooling strategy. When comparing the performance in AIA between **J-F** and **J-G** we can see that the information averaging via visual content is more important than the structure preserva-

Table 5. Experiment results of different settings

Beijing	Precision@20					MRR@IBUR		
Methods	J-P	J-U	J-F	J-G	JEREMIE	J-F	J-G	J-U
e = 1	22.54	26.25	23.62	29.34	**33.34**	33.46	36.52	**46.32**
e = 2	23.76	28.67	24.45	30.66	**34.15**	33.37	36.15	**47.21**
e = 3	23.63	28.68	24.87	31.14	**34.96**	34.21	37.49	**48.37**
e = 4	24.83	29.73	25.36	31.59	**35.48**	35.43	38.37	**48.70**
e = 5	25.48	30.32	26.77	31.71	**36.72**	35.77	38.23	**48.85**
Beijing	Recall@20					MRR@IBPR		
Methods	J-P	J-U	J-F	J-G	JEREMIE	J-F	J-G	J-P
e = 1	58.81	69.56	59.41	76.03	**85.85**	31.64	38.24	**56.34**
e = 2	60.83	74.29	63.17	78.08	**89.80**	33.15	39.95	**56.32**
e = 3	61.85	73.94	62.84	79.10	**90.98**	33.58	41.01	**58.81**
e = 4	62.35	76.02	64.31	80.05	**91.26**	34.65	41.56	**59.63**
e = 5	62.99	77.48	66.22	80.96	**92.96**	34.23	41.77	**59.92**

tion in our semantic feature learning framework. However, the performance in IBUR and IBPR between **J-F** and **J-G** is the opposite of the one in AIA demonstrates that the structure preservation makes greater contribution in both the IBUR and IBPR tasks. This result demonstrates that our proposed model is effective even when the two tasks are not bound together. Combined with the results in all of the three tasks, we can see that our proposed relation structure simultaneously improve the performance of all the three tasks within a unified learning process.

5.6 Parameter Sensitivity Analysis

We evaluate the sensitivity of the parameters in JEREMIE with *Shanghai* dataset in the condition of $e = 5$. We select three significant parameters $\{\alpha, \beta, \theta\}$ to illustrate the MAP@5 and MRR scores. For each parameter to be analysed, we fix other parameters as the original value ($\alpha = 1000, \beta = 10, \theta = 1$). The parameters α, β and θ range from 10^{-3} to 10^3 (also includes 0) for sensitivity test.

All of the curves in Fig. 2(a) keep stable when $\theta \leq 100$, and they all tend to descent from $\theta = 100$ to $\theta = 1000$. While we choose 1 instead of 100 as the optimal value for θ to make our model more flexible in parameter tuning.

In Fig. 2(b) we can see that the MAP scores of AIA are almost stable in a wide range of α. However, the shape of the MRR curves (IBUR and IBPR) show an ascend trend, which means the user interest constraints F_i and F_j is sensible in IBUR and IBPR respectively. To balance the performance, we choose 1000 as the optimal value for α.

Illustrated in Fig. 2(c), the trend of all scores goes up with the increment of β and all of the metrics level off in the condition of $\beta \geq 1$. To obtain a relatively better performance, we choose 10 as the optimal value for β respectively.

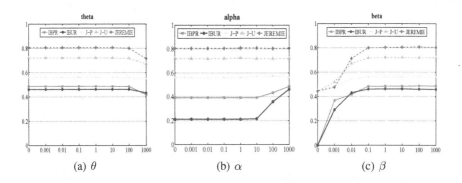

Fig. 2. Sensitivity analysis of θ, α, β in terms of MAP@5 and MRR on the Shanghai dataset

5.7   Space Complexity Analysis

The space requirement for TMC and NMF is $O(n \times m)$. Since TagProp utilizes the KNN technique to reduce the size of search space, its space requirement is $O(n \times K)$. By reducing the problem scale with POI-specific relation matrix partition, the space requirement of our method is $O(n_{max} \times m)$, where n_{max} is the number of images in the largest POI. Typically, $m > 5K$ and $n > 10n_{max}$. Therefore, our method JEREMIE achieves the smallest space requirement than other state-of-the-art models.

5.8   Time Complexity Analysis

For time complexity analysis, we first calculate time complexity for each term in our loss function, then we aggregate them into a final result. The time complexity of each term and corresponding subgradients is listed as:

$$F_i : O(u * m * n_i + u^2 * m)$$
$$G_i : O(d * m * n_i + n_i^2 * m)$$
$$E_i : O(u * m * n_i)$$
$$\nabla_{T_i} A_i : O(u * m * n_i + u * n_i^2 + n_i * u^2 + n_i^2 * m)$$
$$\nabla_{U_i} A_i : O(u * m * n_i + m * u^2)$$
$$\nabla_{W_U} A_i : O(d * n_i^2 * p + d^2 * n_i * p + d^2 * m * p)$$

In our experiment we fix m, d as constants. We approximation u by $u \approx n_k/\omega$, where ω is a constant. The time complexity with user blocks is $O(n_i^3)$. We can obtain similar results by analyzing the symmetric terms of POI blocks. The time complexity of our framework is $O((\max(n_i, n_j))^3)$, where n_i is the number of images in user block i and n_j is the number of images in POI block i.

5.9   Efficiency Analysis

We evaluate the computational efficiency of proposed JEREMIE method in different parallel mode on a 100 K image subset of London. Our algorithm is implemented on MATLAB R2014a, and run on a Intel i7-4770K CPU and 32 GB RAM configured PC. Figure 3 illustrates the running time per iteration of our proposed JEREMIE method in serial and 2-parallel modes. The shape of the curves demonstrates that the parallelism of our proposed JEREMIE method effectively reduces the computational time as the increment of scalability. It is no surprising that our parallel computational framework conducted by matrix partition strategy is the key point of efficiency improvement.

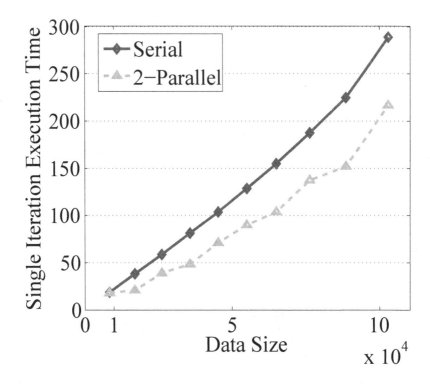

Fig. 3. Serial and parallel execution time of single iteration

6   Conclusion

We propose the JEREMIE method, a Joint SEmantic FeatuRe LEarning model via Multi-relational MatrIx ComplEtion, which jointly complements the semantic features of different entities from heterogeneous domains. We learn the block-wise *image-tag*, *POI-tag* and *user-tag* relations by optimizing a unified objective function which consists of matrix factorization and transductive matrix completion. The overall learning problem can be decomposed into a set of POI-specific and user-specific sub-tasks. Experimental results on automatic image annotation and image-based user (POI) retrieval demonstrate that our approach achieves promising performance on real world social media datasets. In future work, we will fully investigate the connection among users, content, concepts and locations towards a more flexible relation modeling framework.

Acknowledgments. This work was supported in part by the National Natural Science Foundation of China under Grant 61672497, Grant 61332016, Grant 61620106009, Grant 61650202 and Grant U1636214, in part by the National Basic Research Program of China (973 Program) under Grant 2015CB351802, and in part by the Key Research Program of Frontier Sciences of CAS under Grant QYZDJ-SSW-SYS013. This work was also partially supported by CAS Pioneer Hundred Talents Program by Dr. Qiang Qu.

References

1. Cartis, C., Gould, N.I., Toint, P.L.: On the evaluation complexity of composite function minimization with applications to nonconvex nonlinear programming. SIAM J. Optim. **21**(4), 1721–1739 (2011)
2. Chen, M., Zheng, A., Weinberger, K.: Fast image tagging. In: ICML, pp. 1274–1282 (2013)
3. Cho, E., Myers, S.A., Leskovec, J.: Friendship and mobility: user movement in location-based social networks. In: KDD, pp. 1082–1090. ACM (2011)
4. Comaniciu, D., Meer, P.: Mean shift: a robust approach toward feature space analysis. TPAMI **24**(5), 603–619 (2002)
5. Crandall, D.J., Backstrom, L., Huttenlocher, D., Kleinberg, J.: Mapping the world's photos. In: WWW 2009, pp. 761–770. ACM (2009)
6. Guillaumin, M., Mensink, T., Verbeek, J., Schmid, C.: TagProp: discriminative metric learning in nearest neighbor models for image auto-annotation. In: ICCV, pp. 309–316. IEEE (2009)
7. Halpin, H., Robu, V., Shepherd, H.: The complex dynamics of collaborative tagging. In: WWW, pp. 211–220. ACM (2007)
8. Hariharan, B., Zelnik-Manor, L., Varma, M., Vishwanathan, S.: Large scale max-margin multi-label classification with priors. In: ICML, pp. 423–430 (2010)
9. Jenatton, R., Roux, N.L., Bordes, A., Obozinski, G.R.: A latent factor model for highly multi-relational data. In: NIPS, pp. 3167–3175 (2012)
10. Krizhevsky, A., Sutskever, I., Hinton, G.E.: Imagenet classification with deep convolutional neural networks. In: NIPS, pp. 1097–1105 (2012)
11. Lee, D.D., Seung, H.S.: Learning the parts of objects by non-negative matrix factorization. Nature **401**(6755), 788–791 (1999)
12. Li, X., Snoek, C.G., Worring, M.: Learning social tag relevance by neighbor voting. TMM **11**(7), 1310–1322 (2009)
13. Lin, Z., Ding, G., Hu, M., Wang, J., Ye, X.: Image tag completion via image-specific and tag-specific linear sparse reconstructions. In: CVPR, pp. 1618–1625. IEEE (2013)
14. Liu, D., Yan, S., Hua, X.S., Zhang, H.J.: Image retagging using collaborative tag propagation. TMM **13**(4), 702–712 (2011)
15. Liu, D., Ye, G., Chen, C.T., Yan, S., Chang, S.F.: Hybrid social media network. In: ACMMM, pp. 659–668. ACM (2012)
16. Liu, J., Li, Z., Tang, J., Jiang, Y., Lu, H.: Personalized geo-specific tag recommendation for photos on social websites. TMM **16**(3), 588–600 (2014)
17. Mairal, J., Bach, F., Ponce, J., Sapiro, G.: Online learning for matrix factorization and sparse coding. JMLR **11**, 19–60 (2010)
18. Moxley, E., Kleban, J., Manjunath, B.: Spirittagger: a geo-aware tag suggestion tool mined from flickr. In: ACM MIR, pp. 24–30. ACM (2008)
19. Qian, X., Feng, H., Zhao, G., Mei, T.: Personalized recommendation combining user interest and social circle. TKDE **26**(7), 1763–1777 (2014)
20. Russell, B.C., Torralba, A., Murphy, K.P., Freeman, W.T.: LabelMe: a database and web-based tool for image annotation. IJCV **77**(1–3), 157–173 (2008)
21. Singh, A.P., Gordon, G.J.: Relational learning via collective matrix factorization. In: KDD, pp. 650–658. ACM (2008)
22. Sutskever, I., Tenenbaum, J.B., Salakhutdinov, R.R.: Modelling relational data using bayesian clustered tensor factorization. In: NIPS, pp. 1821–1828 (2009)

23. Thomee, B., Shamma, D.A., Friedland, G., Elizalde, B., Ni, K., Poland, D., Borth, D., Li, L.J.: YFCC100M: the new data in multimedia research. Commun. ACM **59**(2), 64–73 (2016)

24. Wang, Z., Zhu, W., Cui, P., Sun, L., Yang, S.: Social media recommendation. In: Ramzan, N., van Zwol, R., Lee, J.S., Clüver, K., Hua, X.S. (eds.) Social Media Retrieval. CCN, pp. 23–42. Springer, London (2013). https://doi.org/10.1007/978-1-4471-4555-4_2

25. Wu, L., Jin, R., Jain, A.K.: Tag completion for image retrieval. TPAMI **35**(3), 716–727 (2013)

26. Wu, L., Yang, L., Yu, N., Hua, X.S.: Learning to tag. In: WWW 2009, pp. 361–370. ACM (2009)

27. Ye, M., Yin, P., Lee, W.C., Lee, D.L.: Exploiting geographical influence for collaborative point-of-interest recommendation. In: SIGIR, pp. 325–334. ACM (2011)

28. Yin, H., Sun, Y., Cui, B., Hu, Z., Chen, L.: LCARS: a location-content-aware recommender system. In: KDD, pp. 221–229. ACM (2013)

29. Zhang, J., Wang, S., Huang, Q.: Location-based parallel tag completion for geo-tagged social image retrieval. In: ICMR, pp. 355–362. ACM (2015)

30. Zhu, G., Yan, S., Ma, Y.: Image tag refinement towards low-rank, content-tag prior and error sparsity. In: ACMMM, pp. 461–470. ACM (2010)

31. Zhu, J.: Max-margin nonparametric latent feature models for link prediction. In: ICML, pp. 719–726 (2012)

A Big Data Driven Approach to Extracting Global Trade Patterns

Giannis Spiliopoulos[1], Dimitrios Zissis[1,2(✉)],
and Konstantinos Chatzikokolakis[1]

[1] MarineTraffic, London, UK
{giannis.spiliopoulos,konstantinos.chatzikokolakis}
@marinetraffic.com
[2] Department of Product and Systems Design Engineering,
University of the Aegean, Syros, Greece
dzissis@aegean.gr

Abstract. Unlike roads, shipping lanes are not carved in stone. Their size, boundaries and content vary over space and time, under the influence of trade and carrier patterns, but also infrastructure investments, climate change, political developments and other complex events. Today we only have a vague understanding of the specific routes vessels follow when travelling between ports, which is an essential metric for calculating any valid maritime statistics and indicators (e.g. trade indicators, emissions and others). Whilst in the past though, maritime surveillance had suffered from a lack of data, current tracking technology has transformed the problem into one of an overabundance of information, as huge amounts of vessel tracking data are slowly becoming available, mostly due to the Automatic Identification System (AIS). Due to the volume of this data, traditional data mining and machine learning approaches are challenged when called upon to decipher the complexity of these environments. In this work, our aim is to transform billions of records of spatiotemporal (AIS) data into information for understanding the patterns of global trade by adopting distributed processing approaches. We describe a four-step approach, which is based on the MapReduce paradigm, and demonstrate its validity in real world conditions.

Keywords: Big spatiotemporal data · AIS · Global shipping routes
K-means clustering · Apache Spark

1 Introduction

Since ancient times, trade has been conducted mostly by sea. Captains of their times sought out the safest, but also fastest routes connecting major trading sea ports. As early as 515 BC, the sailor Scylax, made the first recording of the Mediterranean voyages or sailing instructions, which described safe passages between Mediterranean ports (later known as Hellenic Periploi), listing ports and coastal landmarks with approximate distances and routes between them. Throughout history, vessels have regularly set their shipping courses so as to take advantage of the prevailing winds and ocean currents, leading to the definition of major shipping trade routes, which are

© Springer International Publishing AG 2018
C. Doulkeridis et al. (Eds.): MATES 2017, LNCS 10731, pp. 109–121, 2018.
https://doi.org/10.1007/978-3-319-73521-4_7

mostly in use until today. Such popular trade routes include the routes crossing the Pacific Ocean, the Atlantic Ocean routes and the Indian ocean routes, but unlike roads, these shipping routes are not carved in stone. The size of the corridor, its content and connections, can vary greatly over space and time, under the influence of trade and carrier patterns, but also due to infrastructure investments, climate change, political events and other complex international events. For instance, global warming is having a major effect on shipping routes; as new routes such as the Arctic Ocean shipping route north of Russia (which has cut thousands of miles off the journey from China to the European ports) are emerging [1], while icebergs are disrupting the traditional shipping lanes off the Canadian coast [2]. Similarly, investments in port terminals or canal expansions have widespread effects on routes and trading patterns; such as the recent expansion of the Panama Canal, which influenced a range of stakeholders regarding shipping rates, peripheral port capacity and port investments.

The importance of a well-developed understanding of the maritime traffic patterns and trade routes is critical to all seafarers and stakeholders. From a security perspective, it is necessary for understanding areas of high congestion, so that smaller vessels can avoid collisions with bigger ships. Moreover, an understanding of vessel patterns at scale can assist in the identification of anomalous behaviors and help predict the future location of vessels. Additionally, by combining ship routes with a model to estimate the emission of vessels (which depends on travel distance, speed, draught, weather conditions and characteristics of the vessel itself), emissions of e.g. CO_2 and NOx can be estimated per ship and per national territory [3]. From an economic side, stakeholders selecting to deploy a ship on a particular route need to find the optimal mix between a number of variables such as the shortest path between two ports, cost of route, expected congestion, travel time, size of vessel and capacity and many more. According to the findings of a recent Eurostat funded project, current problems in methods of calculating official maritime statistics include, (i) Distance travelled per ship is now based on an inaccurate average distance matrix for ports, (ii) Missing Information on travel routes for goods to estimate unit prices for transit trade statistics [3].

While in the past, maritime surveillance had suffered from a lack of data, current tracking technology has transformed the problem into one of an overabundance of information. Progressively huge amounts of structured and unstructured data, tracking vessels during their voyages across the seas, are becoming available, mostly due to the Automatic Identification System (AIS) that vessels of specific categories are required to carry. The AIS is a collaborative, self-reporting system that allows marine vessels to broadcast their information to nearby vessels and on-ground base stations. Vessels equipped with AIS transceivers periodically broadcast messages that include the vessel identifying information, characteristics, and destination together with other information coming from on-board equipment, such as current location, speed, and heading. AIS data slowly becoming available, provides almost global coverage as data collection methods are not restricted to a single country or continent, providing an opportunity for in depth analysis of patterns at a global scale which was previously unavailable [3].

However, current Information & Communication Technology (ICT) and traditional data mining approaches are challenged when called upon to decipher the complexity of these environments and produce actionable intelligence. AIS geospatial data-sets are very large in size, containing billions of records, and skewed, as specific regions, can contain substantially more data than others, making processing and storage with conventional methods highly challenging. As such traditional techniques and technologies have proven incapable of dealing with such volumes of loosely structured spatio-temporal data.

In our approach, we exploit a massive volume of historical AIS data to estimate trade routes in a data-driven way, with no reliance on external sources of information. We present a four phased approach which is based on the MapReduce distributed programming paradigm, and demonstrate its effectiveness and validity in real world conditions.

Our work presents novelties on three fronts:

- Distributed computation: We present an architectural prototype which is validated by efficiently processing billions of AIS message (>500 Gb) within a few hours. To the best of our knowledge no previous work has successfully analyzed AIS datasets of this size and coverage within the time scale of our solution (less than 3 h).
- Algorithmic Accuracy: We discuss our algorithmic approach to generating accurate trade routes by overcoming many of the known accuracy issues of AIS in a distributed fashion by adopting a MapReduce approach.
- Domain Specific: We uncover global maritime trade routes which can be used as a method of anomaly detection, investigation but also understanding and predicting variations in trade patterns and the effect of events.

The rest of the manuscript is organized as follows: Sect. 2 shortly presents previous work in this domain, while Sect. 3 describes our approach and Sect. 4 presents the preliminary results. Finally, Sect. 5 presents the conclusion and briefly outlines shortcoming of this work and future improvements.

2   Related Work

AIS data has been used as valid method for extracting valuable information regarding vessel behavior, operational patterns and performance statistics for a number of years now. As Tichavska, Cabrera, Tovar and Arana point out, AIS data has been used for a variety of applications including, optimization of radio propagation channel techniques, real-time statistical processing of traffic information, improving ship traffic management and operations, sustainable transport solutions and many more [4]. Specifically, for route definition and motion pattern extraction, AIS is considered a valid source of data, used as a framework for trajectory forecasting and anomaly detection. Most published works can be categorized per the methods the authors follow which are either (i) grid based or (ii) methods of using vectorial representations of traffic. In (i) grid based approaches, the area of coverage is split into cells which are characterized by the motion properties of the crossing vessels to create a spatial grid. In the second category, vessel trajectories are modeled as a set of connected waypoints. Thus, vessel motions in large areas (e.g., at a global scale) can be managed thanks to the high compactness of the waypoint representation [5, 6].

Towards this direction, in their recent work, Ristic et al. perform statistical analysis of vessel motion patterns to extract motion patterns which are then used to construct the corresponding motion anomaly detectors using adaptive kernel density estimation [7]. Mazzarella et al. apply a Bayesian vessel prediction algorithm based on a Particle Filter (PF) on AIS data [8]. Zhang et al. apply hierarchical and other clustering methods to learn the typical vessel sailing pattern within the waters of Xiamen Bay and Chengsanjiao, China [9]. Pallotta et al. present the TREAD (Traffic Route Extraction and Anomaly Detection) methodology, which relies on the DBSCAN algorithm for automatically detecting anomalies and projecting current trajectories and patterns into the future [10].

As the amount of available AIS data grows to massive scales though, researchers are realising that computational techniques must also contend with acquiring, storing, and processing the data. Applying traditional techniques to AIS data processing can lead to processing times of several days, if applied to global data sets of considerable size. In addition to this, many traditional approaches assume that the underlying data distribution is uniform and spatially continuous. This is not the case for global AIS data, as it is often to have large geographical coverage gaps, message collisions or erroneous messages especially when processing large areas [11, 12]. This problem is mostly evident when dealing with extended geographical areas and "big" datasets. In their majority, previous research efforts have focused on limited geographical areas (e.g. a specific coastal area or sea port) and smaller datasets (e.g. several thousands of AIS messages/GBs) [13–15], often overstepping the problems AIS data quality altogether. In their work [5], authors present a two-step method to achieve a balance between computational time and performance; first performing data simplification by applying the Douglas-Peucker (DP) algorithm before processing the simplified trajectories with Kernel Density Estimation.

Grid-based methods have been considered effective only for small area surveillance and the computational burden was regarded as its limitation when increasing the scale [10]. Therefore a "vectorial" representation of traffic was proposed [16] to allow implementation at a global scale, including waypoint objects and route objects. However, these methods had only been implemented in limited geographical areas (e.g. a 200 × 160 km area in the North Adriatic Sea and similar), and limited information was given about performance [16, 17]. However, in their work Wu et al., demonstrate the ability of a grid-based method for computing shipping density, fast enough to be performed at a global scale (less than 56 h). In this work it took 56 h to produce all the global monthly ship density, traffic density and AIS receiving frequency maps, from August 2012 to April 2015; 33 months of data [17].

To date, very few works, apply the advancements that have been made on the big data front to AIS data processing. In 2008, the MapReduce programming approach was described by Google engineers, in which data-parallel computations are executed on clusters of unreliable machines by systems that automatically provide locality-aware scheduling, fault tolerance, and load balancing while and shortly after in 2011, the Hadoop implementation by Yahoo's engineers was made publicly available under an Apache License [10, 18]. Although certainly not a panacea [19], Hadoop introduced millions of programmers and scientists to parallel and distributed computation, starting the "big data wave". In their work [20], Wang et al. attempt to tackle the big data issue caused by the AIS data for anomaly detection purposes. They implement a two-step process, where they firstly use an unsupervised technique, based upon the

Density-Based Spatial Clustering of Applications with Noise considering Speed and Direction (DBSCAN-SD) incorporating non- spatial attributes, such as speed and direction, to label normal and abnormal position points of vessels based on the raw AIS data. Secondly, they train a supervised learning algorithm designed with the MapReduce paradigm running on Hadoop using the labelled data generated in from the first step. The authors support that the distributed approach is capable of outperforming the promise of traditional GIS applications.

Following Hadoop's success numerous frameworks and open source tools appear on the Big Data ecosystem. Apache Spark, originally designed by researchers at the University of Berkeley, was developed in response to limitations in the MapReduce/Hadoop cluster computing paradigm, which forces a particular linear dataflow structure on distributed programs (acyclic data flow model). Many iterative machine learning algorithms, as well as interactive data analysis tools, reuse a working set of data across multiple parallel operations. Spark processes data in-memory and has been shown to be capable of outperforming Hadoop by $10\times$ in iterative machine learning workloads [18]. In our previous work, we confirmed the potential benefits of applying such techniques to large AIS data processing [21, 22]. In [21, 23] we presented an adaptation of the well-known KDE algorithm to the map-reduce paradigm to estimate a seaports extended area of operation from AIS data. This work is complementary, in that it addresses the problem of estimating global trade routes in an adaptive, scalable, and unsupervised way, based on k-means clustering applied in a distributed fashion. Similarly, [24] Salmon and Ray present their work on designing a hybrid approach based on the Lambda architecture [19] for both real-time and archived data approaches to processing maritime traffic data.

3   Approach

As described in the previous sections, the aim of this work is to calculate the global trade routes from large amounts of AIS data. Out of the 64 different types of AIS messages that can be broadcast by AIS transceivers (as defined by the ITU 1371-4 standard), our work focuses on the 6 most relevant ones, which account for approximately 90% of AIS typical scenarios [19]. Types 1, 2, 3, 18, and 19 are position reports, which include latitude, longitude, speed-over-ground (SOG), course-over-ground (COG), and other fields related to ship movement; type 5 messages contain static-and voyage information, which includes the IMO identifier, radio call sign, name, ship dimensions, ship and cargo types. In all messages, each vessel is identified by its Marine Mobile Service Identifier (MMSI) number. Data is received through the MarineTraffic system and for the purposes of this work we use a dataset of approximately 5 billion messages (i.e., 525 GB) recorded from January to December 2016 (Table 1). In the rest of this section we present the methodology followed to transform the raw data collected into meaningful information and thus useful for data analysis. We provide a detailed analysis of

Table 1. Original dataset statistics

Dataset statistics	
Time period	January–December 2016
Positions count	>5 Billion
Port calls count	>3 Million
Number of unique vessels	>200K
Ports covered in dataset	>3K

each step of the approach and explain thoroughly the effect our actions had to the considered dataset.

3.1 Distributed Processing

For the distributed processing tasks, we rely on a HDInsight Azure Spark (2.1.0 version) cluster made up by: 6 worker nodes (D4v2 Azure nodes), each one equipped with 8 processing cores and 28 GB RAM; and 2 head nodes (D12 v2 Azure nodes), each

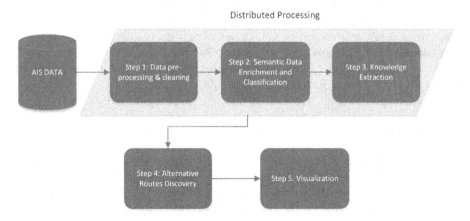

Fig. 1. Four step distributed approach

one equipped with 4 processing cores and 28 GB RAM, summing up to a total of 56 computing cores and 224 GB RAM. This setup has been sufficient to cover the processing requirements of all the complex computations of our methodology (Fig. 1).

3.2 Step 1: Trajectory Data Cleaning and Preprocessing

AIS messages do not provide any trustworthy voyage information with respect to departure and destination ports. In fact, the only relevant data collected from AIS are type 5 messages which include the destination port information. However, this is manually given by the ship's crew and prone to errors, inconsistencies etc. Thus, it is fundamental to discover such knowledge (i.e., departure and destination ports) from the AIS positional

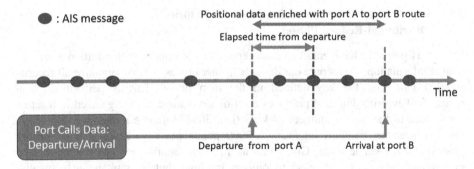

Fig. 2. AIS message data enrichment

data to perform route analysis. We algorithmically, calculate port destination and port departure for each AIS message on spark as depicted in Fig. 2 below. The purpose of such correlation is to assign accurately the departure port, the destination port based on the timestamp of each AIS message and the departure time and arrival time respectively.

In addition to route assignment, for each message we compute the time elapsed (measured in minutes) since the vessel's departure based on the reported timestamps (i.e. from AIS messages). The time elapsed field enables us to include the time dimension in our analysis on AIS messages and group them into time-aligned routes with the same <departure, destination> port and vessel type.

We focus our work only on AIS messages originating from cargo and tanker vessels, as we are not interested in smaller vessels such as fishing boats, tugs, etc. that do not follow a repetitive pattern and may not be representative of global trade routes. Similarly, for this study, we filter out AIS messages originating from passenger vessels as these follow different patterns e.g. visiting different ports than cargo vessels and messages with recorded speed less than 0.5 knots which is considered as the lower bound for vessels moving underway using their engines. The resulting dataset of this preprocessing steps results in approximately 1 billion enriched AIS messages and approximately 28K distinct vessels.

3.3 Step 2: Semantic Data Enrichment and Classification (Distributed-Map Phase)

The preprocessing step uses a dataset of approximately $3 * 10^3$ ports resulting in a large number of port-to-port combinations (i.e., practically almost any port may be connected with any other port in the world resulting in $9 * 10^6$ combinations), which are further increased as we take into account also the ship-type. However, it makes sense to apply any algorithmic approach on data on each route separately. Thus, in order to take advantage of the parallelization ability of spark we map each enriched message to a key-value pair. The key uniquely identifies the route per vessel type and it is generated as the concatenated unique identifier of departure – destination pair and the vessel type, while the value is the enriched message itself.

3.4 Step 3: Knowledge Extraction from Trajectories (Distributed-Reduce Phase)

Up to this point we have enriched each record of our dataset with additional voyage related information. To prepare our data for further processing, we organize all records on lists based on the key defined in the map phase. This is performed by a reduce-by-key procedure and produces a set of key-valued pairs organized in a set of rows equal to the distinct number of keys (i.e., 368473 unique routes per ship type). Each reduced set contains on average 2000 points per key, which can be further processed by a single node. Given that at this time, spark does not support nested map-reduce processes, we select to process multiple routes simultaneously by distributing their keys to multiple nodes, instead of splitting each set computation to multiple nodes. In order to capture a unique route for all the points included in each set we perform the well-known k-means clustering technique using WEKA, an open source software collection of machine learning algorithms for data mining tasks. K-means is employed to cluster raw points and the features selected for this process are the following:

- latitude,
- longitude,
- relative timestamp.

This enables our solution to detect clusters of vessels' positions based on both location and time with respect to the departure timestamp, enhancing route perception with average elapsed time (or within a time range) from departure for vessels of the same type with the same route that sail in close by trajectories.

The number of clusters per route K is selected dynamically using the formula below.

$$K = \max\left(\min\left(\left(\frac{N}{10}\right), cmax\right), 1\right)$$

where,

- cmax is the maximum number of clusters (set to 100 for our evaluation),
- N is the #points per route.

The arbitrary selection of the maximum number of centroids cmax is equal to 100. The cmax selection is based on the degree of compression of information that we would like to achieve, while preserving some of its original granularity for the global dataset. Finally, before moving to the next step, we assign to each centroid the number of distinct vessels that have at least one point within the cluster.

3.5 Step 4: Discovering Alternative Routes

During the previous steps, it was assumed that same vessel types sailing on the same route will follow similar trajectories. However, this is not always accurate, as various factors such as weather conditions, draught, etc. may vastly differentiate the vessel's course. In such cases we observed that the outcome of the clustering phase suffered from fluctuating events and the produced trajectories had continuous changes of vessel's course. In most cases the actual data indicated the existence of (at least) two different courses for the same route. To identify these courses, we applied a trajectory splitting algorithm based on the following predefined set of rules.

We examine all route points using a three points sliding observation window. Each point is assigned to a different course if any of the empirical rules evaluated in the following algorithm is valid.

For each course of a route detected do the following:

For each three sequential points p_{i-1}, p_i, p_{i+1} evaluate the following rules:

1. $abs(\theta_{i-1,i} - \theta_{i,i+1}) > 35^{\circ}$
2. $D_{i-1,i} + D_{i,i+1} > 1.5 * D_{i-1,i+1}$
3. $D_{i-1,i} + D_{i,i+1} > \frac{2*D_{1,N}}{N}$
4. $V_{i-1,i} > V_{th}$

where,

- $\theta_{i,j}$ is the bearing angle between p_i, p_j,
- $D_{i,j}$ is the distance between p_i, p_j,
- $V_{i,j}$ is the vessel's speed so as to reach p_j from p_i,
- V_{th} is the speed threshold, which is set to 40 knots, as cargo ships and tankers typically have speed less than 35 knots,
- N is the #points per route.

Although multiple sailing courses are linked with the same route, they are completely different paths from departure port to destination port, and thus, when visualizing results we treat each sailing course as a different route. In the following section we analyze the evaluation results of our methodology and provide insights on some interesting new routes such as China to east USA.

4 Results

In this section, we present some preliminary results of the approach described in the previous section. All results presented below have been produced using the cluster setup presented in Sect. 3.1 and the execution times correspond to the results of the spark processing with 24 executors having two cores each and kryo serializer option enabled to minimize the serialization cost.

As previously discussed in Sect. 3 our approach is a four stepped approach. In the preprocess step, we parsed the original dataset of 5 billion records and filtered out the messages that do not originate from cargo or tanker vessels based on the AIS vessel type recorded. Then, positional data were combined with recorded departures and destinations to create a smaller dataset with enriched data of 0.79 billion records. Processing efforts were split into 5 jobs, 12 stages and 3727 tasks, having a total computational cost of 1.3 h. The map phase was executed through 1 job, 1 stage and 267 tasks and lasted almost 10 min. The reduce phase including the splitting process was executed in 1 jobs, 1 stage and 267 tasks and had a 10 min computational cost. The entire process from the initial import of raw data to end results extraction lasted approximately two hours.

The resulting dataset was stored in a single 850 MB csv file including information for all the vessels. The total number of routes detected, after splitting was 440,854 represented with 10,847,328 points (Fig. 3). The result represents a significant part of information in terms of routes detected, i.e. the number of routes detected is greater

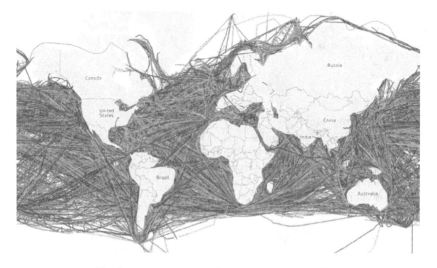

Fig. 3. Routes extracted for cargo vessels for 2016.

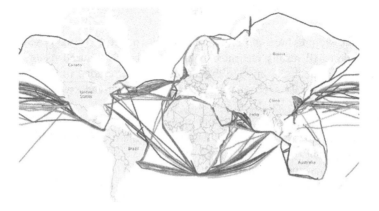

Fig. 4. Overview of the routes connection some of the top 50 ports around the globe.

than the distinct number of routes per vessel type, while the compression rate is 73:1 in terms of positions. It should be noted that since the dataset used is recorded from January to December 2016, some vessel routes may start or end in the middle of the sea because the ship's voyage is not entirely in this period (i.e. ship's voyage may start before 2016, or may end in 2017).

Through our approach, we have been able to control the degree of compression of the resulting dataset based on the selection of K, being capable to create visualizations of sets of global routes. Regarding the domain, it is interesting to view that we can validate most popular trade routes with accuracy and new routes are uncovered, such as the artic route above Russia and new routes that pass over North America to link east USA with China have been detected (Fig. 4). Finally, the addition of time in the clustering process for K-means introduces an average voyage time estimation on each centroid detected, this estimate could be used a baseline for future works on optimal route selection for the maritime industry.

5 Conclusion and Future Work

This article focused on the challenges of analyzing huge amounts of vessel tracking data produced through the AIS. The novelty of the method is in the direction of adopting a map reduce approach to distribute the computational burden across a cluster of commodity machines to perform the computation in approximately 2 h time. Preliminary results presented in the previous section confirm the validity of the adopted approach. Future work, will be focused on improvements of the algorithmic approach to improve the accuracy of the identified routes.

Acknowledgement. This project has received funding from the European Union's Horizon 2020 research and innovation programme under grant agreement No. 732310 and by Microsoft Research through a Microsoft Azure for Research Award.

References

1. Savadove, B.: China Begins Using Arctic Shipping Route That Could "Change the Face of World Trade," Bus. Insid. (2013)
2. Euronews: Arctic icebergs reach Canadian coast, disrupting sea lanes and fishing. Euronews (2017)
3. Tessa de Wit, A., Consten, M., Puts, C., Pierrakou, M., Bis, A., Bilska, et al.: ESSnet Big Data Deliverable 4.2 Deriving port visits and linking data from Maritime statistics with AIS-data (2017)
4. Cabrera, F., Molina, N., Tichavska, M., Arana, M.: Design of a low cost prototype of Automatic Identification System (AIS) receiver. In: 2015 1st URSI Atlantic Radio Science Conference (URSI AT-RASC), p. 1. IEEE (2015). https://doi.org/10.1109/URSI-AT-RASC.2015.7303000

5. Li, Y., Liu, R.W., Liu, J., Huang, Y., Hu, B., Wang, K.: Trajectory compression-guided visualization of spatio-temporal AIS vessel density. In: 2016 8th International Conference on Wireless Communication Signal Process, pp. 1–5. IEEE (2016). https://doi.org/10.1109/WCSP.2016.7752733
6. Fiorini, M., Capata, A., Bloisi, D.D.: AIS data visualization for Maritime Spatial Planning (MSP). Int. J. E-Navig. Marit. Econ. 5, 45–60 (2016). https://doi.org/10.1016/j.enavi.2016.12.004
7. Ristic, B., La Scala, B., Morelande, M., Gordon, N.: Statistical analysis of motion patterns in AIS data: anomaly detection and motion prediction. IEEE (2008)
8. Mazzarella, F., Arguedas, V.F., Vespe, M.: Knowledge-based vessel position prediction using historical AIS data. In: 2015 Sensor Data Fusion: Trends, Solutions, Applications, pp. 1–6. IEEE (2015). https://doi.org/10.1109/SDF.2015.7347707
9. Zhang, W., Goerlandt, F., Kujala, P., Wang, Y., Nikitakos, N.: An advanced method for detecting possible near miss ship collisions from AIS data. Ocean Eng. 124, 141–156 (2016). https://doi.org/10.1016/j.oceaneng.2016.07.059
10. Pallotta, G., Vespe, M., Bryan, K.: Vessel pattern knowledge discovery from AIS data: a framework for anomaly detection and route prediction. Entropy 15, 2218–2245 (2013). https://doi.org/10.3390/e15062218
11. Poļevskis, J., Krastiņš, M., Korāts, G., Skorodumovs, A., Trokšs, J.: Methods for processing and interpretation of AIS signals corrupted by noise and packet collisions. Latv. J. Phys. Tech. Sci. 49, 25–31 (2012). https://doi.org/10.2478/v10047-012-0015-3
12. Yang, M., Zou, Y., Fang, L.: Collision and detection performance with three overlap signal collisions in space-based AIS reception. In: 2012 IEEE 11th International Conference on Trust, Security and Privacy in Computing and Communications, pp. 1641–1648. IEEE (2012). https://doi.org/10.1109/TrustCom.2012.109
13. Willems, N., van de Wetering, H., van Wijk, J.J.: Visualization of vessel movements. Comput. Graph. Forum. 28, 959–966 (2009). https://doi.org/10.1111/j.1467-8659.2009.01440.x
14. Willems, N., van de Wetering, H., van Wijk, J.J.: Evaluation of the visibility of vessel movement features in trajectory visualizations. Comput. Graph. Forum 30, 801–810 (2011). https://doi.org/10.1111/j.1467-8659.2011.01929.x
15. Di Battista, G., Fekete, J.-D., Qu, H.: Technical Committee on Visualization and Graphics, Proceedings of IEEE Pacific Visualization Symposium 2011, Hong Kong, China, 1–4 March 2011. IEEE Computer Society, Institute of Electrical and Electronics Engineers (2011)
16. Arguedas, V., Vespe, M., Pallotta, G.: Automatic generation of geographical networks for maritime traffic surveillance. In: 2014 17th International Conference on Information Fusion (FUSION). IEEE (2014)
17. Wu, L., Xu, Y., Wang, Q., Wang, F., Xu, Z.: Mapping global shipping density from AIS data. J. Navig. 70, 67–81 (2017). https://doi.org/10.1017/S0373463316000345
18. Zaharia, M., Chowdhury, M., Franklin, M.J., Shenker, S., Stoica, I.: Spark: cluster computing with working sets (2010)
19. Stonebraker, M.: Hadoop at a Crossroads? n.d.
20. Wang, X., Liu, X., Liu, B., de Souza, E.N., Matwin, S.: Vessel route anomaly detection with Hadoop MapReduce. In: 2014 IEEE International Conference on Big Data (Big Data), pp. 25–30. IEEE (2014). https://doi.org/10.1109/BigData.2014.7004464
21. Millefiori, L.M., Zissis, D., Cazzanti, L., Arcieri, G.: A distributed approach to estimating sea port operational regions from lots of AIS data. In: 2016 IEEE International Conference on Big Data (Big Data), pp. 1627–1632. IEEE (2016). https://doi.org/10.1109/BigData.2016.7840774

22. Millefiori, L., Zissis, D., Cazzanti, L., Arcieri, G.: Computational Maritime Situational Awareness Techniques for Unsupervised Port Area, Nato Unclassified Reports, Science and Technology Organization Centre for Maritime Research and Experimentation, La Spezia, Italy (2016)
23. Millefiori, L.M., Zissis, D., Cazzanti, L., Arcieri, G.: Scalable and distributed sea port operational areas estimation from AIS data. In: 2016 IEEE 16th International Conference on Data Mining Workshops, pp. 374–381. IEEE (2016). https://doi.org/10.1109/ICDMW.2016.0060
24. Salmon, L., Ray, C.: Design principles of a stream-based framework for mobility analysis. Geoinformatica **21**, 237–261 (2017). https://doi.org/10.1007/s10707-016-0256-z

Efficient Processing of Spatiotemporal Pattern Queries on Historical Frequent Co-Movement Pattern Datasets

Shahab Helmi[✉] and Farnoush Banaei-Kashani[✉]

Department of Computer Science and Engineering,
University of Colorado Denver, Denver, USA
{shahab.helmi,farnoush.banaei-kashani}@ucdenver.edu

Abstract. Thanks to recent prevalence of location sensors, collecting massive spatiotemporal datasets containing moving object trajectories has become possible, providing an exceptional opportunity to derive interesting insights about behavior of the moving objects such as people, animals, and vehicles. In particular, mining patterns from co-movements of objects (such as movements by players of a sports team, joints of a person while walking, and cars in a transportation network) can lead to the discovery of interesting patterns (e.g., offense tactics of a sports team, gait signature of a person, and driving behaviors causing heavy traffic). Given a dataset of frequent co-movement patterns, various spatial and spatiotemporal queries can be posed to retrieve relevant patterns among all generated patterns from the pattern dataset. We term such queries, *pattern queries*. Co-movement patterns are often numerous due to combinatorial complexity of such patterns, and therefore, co-movement pattern datasets often grow very large in size, rendering naive execution of the pattern queries ineffective. In this paper, we propose the *FCPIR framework*, which offers a variety of index structures for efficient answering of various range pattern queries on massive co-movement pattern datasets, namely, spatial range pattern queries, temporal range (time-slice) pattern queries, and spatiotemporal range pattern queries.

Keywords: Spatiotemporal indexing
Spatiotemporal query processing · Pattern query

1 Introduction

Recent advances in location sensors, for instance wearable devices and cell phones with built-in GPS sensors, have enabled collection of massive moving object datasets. We term these datasets multi-variate spatiotemporal event sequence datasets (or MVS datasets, for short) where each event captures the location of an object (among other possible information) at a specific time, and accordingly, each event sequence (or variate) represents the trajectory of an object. MVSs can capture movements/events about various objects in different applications; e.g., joints in human body as a person walks, vehicles navigating a transportation

© Springer International Publishing AG 2018
C. Doulkeridis et al. (Eds.): MATES 2017, LNCS 10731, pp. 122–137, 2018.
https://doi.org/10.1007/978-3-319-73521-4_8

network, ants in a colony as they follow their daily affairs, players in sports teams (e.g., soccer teams) competing on a field, etc. In this paper, we are interested in frequent co-movement patterns, where a frequent co-movement pattern is a pattern of movement for a set of objects that are temporally (not necessarily spatially) close and repeatedly appears in various time windows. Here, we describe three use-cases of frequent co-movement pattern mining[1]: (i) in traffic behavior analysis, finding frequent co-movement patterns of vehicles on a transportation network enables study of the dominating driving behaviors that determine and affect traffic condition in the network; (ii) in human gait analysis, skeletal gait of human, where movement of each join is captured by a variate in the corresponding MVS, can be analyzed to identify joint co-movement patterns that uniquely identify a gait, e.g., for gait-based identification; (iii) in sports analytics, tactics of sports teams can be uncovered by discovering frequent co-movement patterns from MVS datasets collected from players during games.

Co-movement patterns are often numerous due to combinatorial complexity of such patterns, and therefore, the mined co-movement pattern datasets often grow very large in size. Accordingly, naive exploration of such pattern datasets to retrieve relevant patterns becomes very time-consuming, if not infeasible. We term such search queries to retrieve relevant patterns from co-movement pattern datasets, co-movement pattern queries (or *pattern queries* for short). Various conventional (spatiotemporal) queries can be posed as pattern queries. In this paper, we propose the *Frequent Co-movement Pattern Query Indexing and Retrieval (FCPIR)* framework, which introduces novel index structures and query processing algorithms, in order to efficiently process pattern queries on *frequent co-movement pattern datasets*. In particular, we present how *FCPIR* enables efficient processing of three spatiotemporal query families on such frequent pattern datasets: spatial range pattern queries, temporal (or time-slice) range pattern queries, and spatiotemporal range pattern queries.

Such queries are among main types of queries used for exploration of pattern datasets. For example in traffic behavior analysis, one can find and compare frequent driving patterns in the downtown versus suburbs using spatial range pattern queries, while the frequent driving patterns in the whole city can be studied during the rush hours versus regular hours using temporal range pattern queries. Finally, spatiotemporal range pattern query can be used to study frequent driving patterns that have caused an accident chain on a major highway.

The remainder of this paper is organized as follows. Section 2 reviews the related work. Section 3 presents necessary notations and formal definition of the problem. In Sect. 4, we present the *SFCMP-tree* and *LM* query processing algorithm for spatial range pattern queries. In Sect. 5, the *TFCMP-tree* and the *TSPQ* query processing algorithms are proposed for temporal range pattern queries. The *LM-MIF* query processing algorithm is proposed for spatiotemporal range pattern queries in Sect. 6. Experimental results are provided in Sect. 7, and Sect. 8 concludes the paper and discusses the future work.

---

[1] See [2] for more details about existing methods for MVS frequent co-movement pattern mining.

2   Related Work

In this section, we review the most relevant bodies of literature to our work. Section 2.1 presents trajectory indexing approaches, and Sect. 2.2 reviews indexing of frequent patterns. Finally, mobility pattern queries are reviewed in Sect. 2.3

2.1   Trajectory Indexing and Retrieval

Spatiotemporal range queries are used to retrieve the trajectories that are contained in (or intersect with) a given spatiotemporal range. Index structures proposed for these types of queries can be roughly categorized into three groups. The first group treats the temporal dimension similar to the spatial dimensions by adding it to the traditional 2D R-trees (e.g., STR-tree [9].) However, the performance of 3D R-trees significantly degrades if the given dataset is dynamic or trajectories in the dataset are long. This is due to exponential increase in the number of conflicts between MBRs in such 3D R-trees. In order to reduce the number of confilcts, the second group of work proposes constructing a separate R-tree at each time-stamp (time-interval) for the part of dataset introduced in that time-stamp (e.g., HR$^+$-tree [11]). With the third group, the region of interest is divided into a number of cells. Each cell stores the information of those trajectory segments that intersect it (e.g., SETI [1]).

In this paper, we study pattern range queries to retrieve co-movement patterns from pattern datasets, rather than range queries on trajectory datasets. Moreover, we are interested in the retrieval of frequent patterns; hence, checking for intersecting patterns in a given range is not sufficient and the number (frequency) of range intersection must be considered as well.

2.2   Mobility Pattern Queries on Trajectory Datasets

In [6], authors propose a query language for mobility patterns. Similarly, in [10] generic operators are described to support retrieval of known mobility queries (such as finding all trajectories that have moved from point a to point b and finally stayed in point c for 10 min). However, since we do not assume any predefined constraints for co-movement patterns, to find all frequent co-movement patterns using the aforementioned methods we need to generate all possible co-movement patterns to be processed using methods such as those discussed in [10], which is most inefficient if at all feasible. To address this problem, we introduce a two-step pattern mining approach instead. With this approach, first we mine all frequent co-movement patterns from a given trajectory dataset using a spatiotemporal extension of the well-known apriori algorithm that we proposed in our prior work [2]. In the second step, which is the focus of this work, we introduce pattern index structures and corresponding query processing algorithms to efficiently retrieve such patterns from frequent co-movement pattern datasets.

2.3   Frequent Pattern Indexing and Management

Indexing structures for frequent patterns can be categorized into two major groups. The first group focuses on similarity search, e.g., finding the k-nearest patterns to a given pattern (e.g., see [7]), while the second group aims at compressing frequent pattern datasets, e.g., by storing only maximal [12] or closed [8] patterns. However, they are neither developed for trajectory (spatial) data, nor allow time-slice (temporal) pattern queries which should account for cases where some frequent patterns become infrequent during a given time-slice. The closest work to our work is [4], where authors propose an index structure for spatiotemporal range queries by repurposing the traditional R-trees in order to reduce the number of disk accesses. However, this approach is only applicable to periodic patterns. Moreover, it only considers spatiotemporal intersection of patterns with the given range not the frequency of intersection. Finally, it only returns the id of qualified objects not the actual trajectories, while in this paper we are interested in retrieving the actual patterns not just their ids.

3   Problem Definition

Let $O = \{O_1, O_2, \ldots, O_n\}$ be a set of n moving objects. For example, in Fig. 1 $O = \{object1, object2, object3\}$. Let S_i show the trajectory of object O_i denoting the spatial path that it follows over time. S_i can be captured as a sequence of locations in 2D space ordered by time as follows: $S_i = (l_{t_1}, l_{t_2}, \ldots, l_{t_m})$, where l_{t_j} shows the location of O_i at the j_{th} timestamp. In this paper, frequent co-movement patterns are of interest, where a frequent co-movement pattern is a pattern of movement for a subset of objects in O that repeatedly appears in various time windows. In this paper, we assume a discrete spatial model (i.e., a grid) by dividing the entire region of interest into smaller regions and replacing absolute locations of objects by their corresponding regions. For example, in Fig. 1, the trajectory of object 1 can be modeled as $S_1 = (WA_{t_1}, ID_{t_2}, MT_{t_3}, ID_{t_4}, WY_{t_5})$. Finally, we define multivariate spatial event sequence $MVS = \{S_1, S_2, \ldots, S_n\}$ as a set of n moving object trajectories. In the rest of this section, we provide necessary definitions to define pattern queries on a given dataset of frequent co-movement patterns, which is assumed to be (previously) mined from an MVS dataset.

Definition 1 (Co-Movement Pattern). *We define a co-movement pattern as $\alpha = \{s_1, s_2, \ldots\}$, where each s_i is a continuous sub-sequence of a trajectory S_i. $t_s(\alpha)$ and $t_e(\alpha)$ are also defined to represent the start time and end time of the co-movement pattern α, respectively. We also define the time-span of the co-movement pattern α as $T(\alpha) = (t_e(\alpha) - t_s(\alpha)) + 1$.*

Definition 2 (Valid Co-Movement Pattern). *We consider a co-movement pattern α as a valid co-movement pattern if it satisfies the following conditions: (i) α cannot be empty (ii) α must not contain more than one sub-sequence s_i from the same event sequence S_i, (iii) all sub-sequences in α must have the*

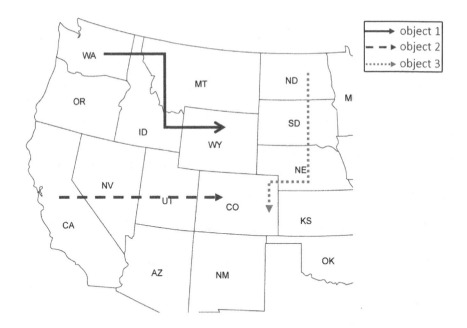

Fig. 1. A co-movement pattern including three objects

same length, i.e., the same number of regions in their sequence, and (iv) $T(\alpha) \leq T_{max}$, where T_{max} is user-defined maximum allowed time-span of a co-movement pattern.

To elaborate, Condition i is self-explanatory; Condition ii ensures we only consider patterns across different objects; Condition iii limits the patterns to those that include same number of events from each object invoked in the pattern; and finally, Condition iv ensures we only consider temporally local patterns i.e., patterns that appear in the same time window rather than far apart in time. For the sake of simplicity, in the rest of this paper we use co-movement pattern and valid co-movement pattern interchangeably.

In Fig. 1, $\alpha = \{(ID, MT, ID)_{S_1}, (CA, NV, UT)_{S_2}\}$ is a valid patterns with $t_s(\alpha) = 1$, $t_e(\alpha) = 4$, and $T(\alpha) = (4-1) + 1 = 4$.

Definition 3 (Sub-Pattern and Super-Pattern). *We say α is a sub-pattern of the co-movement pattern α', i.e., $\alpha \subset \alpha'$ (or α' is a pattern of the α) if for each $s_i \in \alpha$ there is a $s'_i \in \alpha'$, such that s_i is a continuous sub-sequence of s'_i.*

In Fig. 1, if $\alpha = \{(ID, MT)_{S_1}\}$, $\alpha' = \{(ID, MT, ID)_{S_1}\}$, and $\alpha'' = \{(ID, MT, ID)_{S_1}, (CA, NV, UT)_{S_2}\}$, then $\alpha \subset \alpha' \subset \alpha''$.

Definition 4 (Occurrence and Support). *We say an occurrence of co-movement pattern $\alpha = \{s_1, s_2, \ldots, s_k\}$ exists in a multivariate spatial event sequence dataset MVS, if MVS contains α. We denote all occurrences of α in an MVS as $occ(\alpha) = (\lambda_1, \lambda_2, ..., \lambda_f)$, where λ_j captures the start and end time* *of each occurrence, and* $sup(\alpha) = |occ(\alpha)|$ *denotes the support of a co-movement pattern* α *in MVS dataset.*

Note that our proposed framework allows for any of the numerous existing frequency measures to count occurrences of a pattern in sequence data (e.g., [3]) assuming they preserve the monotonicity property, (i.e., the frequency of a co-movement pattern α cannot be greater than those of its sub-patterns).
  In Fig. 1, for $\alpha = \{(ID, WY)_{S_1}\}$, $occ(\alpha) = ([4,5])$ and $|occ(\alpha)| = 1$.

Definition 5 (Frequent Co-Movement Pattern). *A co-movement pattern* α *is a frequent co-movement pattern, if* $sup(\alpha) \geq \mu$, *where* μ *is the user-defined minimum support threshold. Given a maximum time-span* T_{max}, *a minimum-support* μ, *and a multivariate spatial event sequence dataset MVS of n moving objects, we assume F denotes the set of all frequent co-movement patterns mined from the given MVS. Accordingly, we define FCPD (short for frequent co-movement pattern dataset) a pattern dataset containing F, where for each pattern* α *in F we store* $id(\alpha)$, *which is unique for each pattern, as well as* $occ(\alpha)$ *along with the pattern itself.*

Given the above definitions, we define the patterns queries studied in this paper as follows.

Definition 6 (Spatial Range Pattern Query (SRPQ)). *Given an FCPD and a spatial range query* Q_r *containing the set of grid regions* $Q_r = \{r_1, r_2, ...\}$ *from the grid superimposed on the FCPD's region of interest, a spatial range pattern query* Q_S *returns a subset* $P \subseteq FCPD$ *of frequent co-movement patterns in the given FCPD such that each pattern* $\alpha \in P$ *is fully contained in* Q_r, *i.e., all events in* α *are located inside some* $r_i \in Q_r$.

Definition 7 (Temporal Range (or Time-Slice) Pattern Query (TRPQ)). *Given an FCPD and a temporal range query (time-slice)* Q_t *which is a continuous time-interval* $Q_t = [t_s, t_e]$, *a temporal range pattern query* Q_T *returns a subset* $P \subseteq FCPD$ *of frequent co-movement patterns in the given FCPD such that each pattern* $\alpha \in P$ *frequently occurs during* Q_t, *i.e., for each pattern* $\alpha \in P$, $sup(\alpha) \geq \mu$ *during* Q_t.

Definition 8 (Spatiotemporal Range Pattern Query (STRPQ)). *Given an FCPD and a spatial range query* Q_r, *and a temporal range query* Q_t, *a spatiotemporal range pattern query* $Q_{ST} = < Q_r, Q_t >$ *returns a subset* $P \subseteq FCPD$ *of frequent co-movement patterns in the given FCPD such that each pattern* $\alpha \in P$ *is fully contained in* Q_r *and frequently occurs during* Q_t.

In Sects. 4 through 6, we propose efficient index structures and query processing algorithms for each of the aforementioned queries, respectively.

4   Spatial Range Pattern Queries (SRPQ)

In this section, first we present the *SFCMP-tree* (short for Spatial Frequent Co-Movement Pattern), a clustered index structure for efficient *SRPQ* processing. Then, we describe our *Longest-Match (LM)* query processing algorithm which exploits the *SFCMP-tree* in order to efficiently retrieve qualified frequent co-movement patterns for a given query Q_S.

4.1   SFCMP Tree

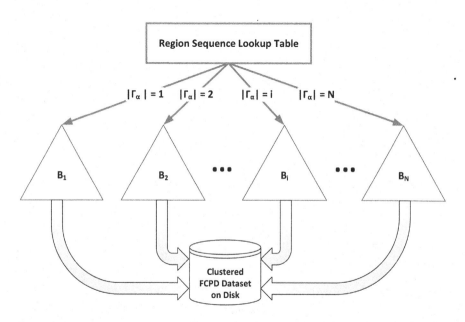

Fig. 2. High-level structure of the *SFCMP-tree* for SRPQ processing

Definition 9 (Region Sequence). *Suppose C is the set of regions covered by α, i.e., the set of regions where each region is the location of at least one of the events in α. We define the region sequence of α, $\Gamma(\alpha)$, as the ordered set of regions in C, where the order is arbitrary, e.g., based on lexicographic order of IDs associated with the regions. We also define $|\Gamma(\alpha)|$ as the level of α.*

For example, if $\alpha = \{(ID, MT, ID)_{S_1}\}$, and $\alpha' = \{(ID, MT, ID)_{S_1}, (CA, NV, UT)_{S_2}\}$, $\Gamma(\alpha) = (ID, MT)$ and $\Gamma(\alpha)' = (CA, ID, MT, NV, UT)$.

Figure 2 shows the high-level structure of the *SFCMP-tree*, τ_s. The *SFCMP-tree* consists of a collection of B^+-tree index structures. Each B^+-tree B_i indexes patterns $\alpha \in FCPD$ that are at level i (i.e., $|\Gamma(\alpha)| = i$), for $i = 1 \cdots N$, where N is the highest level. The region sequence of the pattern α is used as the index key to construct the B^+-tree indexes. Moreover, the B^+-tree indexes are clustered index structures. This is ensured by ordering patterns in the given $FCPD$ first based on their level, and within each level, based on the corresponding region sequence of the patterns at that level. The ordered dataset is then stored in the disk as a clustered file.

4.2   LM Query Processing Algorithm

Lemma 1 (Apriori Property). *If all regions in a region sequence Γ_i for a frequent co-movement pattern α are contained in the given query Q_r, then each pattern α' with $\Gamma_i' \subset \Gamma_i$ is also contained in Q_r.*

With the *Largest-Match(LM) SRPQ* processing algorithm, once query Q_r is received, first the region sequence and level of the pattern query is generated as defined in Sect. 4.1. Next, the region sequence lookup table of *SFCMP-tree* is used to identify the B^+-tree index B_i which indexes patterns at the same level. Thereafter, B_i is traversed using the region sequences of the Q_r as the search key to identify the leaf node that contains the same region sequence (if it exists in $FCPD$). This leaf node has a pointer to a page/block on the disk where all patterns with the same region sequence are clustered; these patterns will be efficiently retrieved to be added to the result set P. However, note that given Lemma 1, we know that all patterns whose regions sequences are sub-sequences of the regions sequence Q_r also belong to P. To retrieve such patterns, the naive approach is to recursively generate all sub-sequences of the Q_r region sequence, and run them as separate queries (as explained above), adding their result to P. As mentioned in Sect. 4.1, with *SFCMP-tree* we include pointers in the file itself that directly point to the pattern groups with regions sequences Γ_i's, where Γ_i's are sub-sequences of the region sequence Γ_i. Therefore, with *SFCMP-tree* we can directly retrieve all relevant patterns without having to traverse the tree for each sub-query.

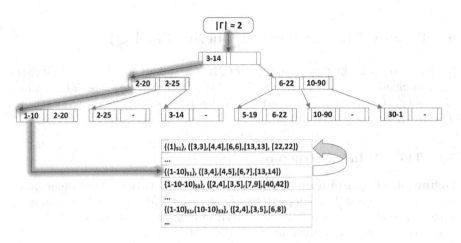

Fig. 3. A detailed example of SFCMP-tree

As an example, consider $Q_r = \{1, 10\}$. the region sequence of this level 2 Q_r is $1 - 10$. Figure 3 (partially) shows the path which will be traversed to retrieve patterns to process Q_r.

Algorithm 1 illustrates the pseudocode for *LM* algorithm. The input to the algorithm is a set of regions Q_{r_0}, and the output of the algorithm is P, which contains all frequent co-movement patterns that are completely inside Q_{r_0}. First result set P and query queue q are initialized to $\emptyset$ and Q_{r_0}, respectively (Line 3). Then if q is not empty, the head element of q is popped into a temporarily variable *tmp*. Thereafter, the *Search* method is called to check if τ_s contains *tmp*, and if so, the corresponding node is returned and added to p (Lines 5–6). Then, using the pointer of p all patterns that their region sequences are equal to that of Q_{r_i}, are fetched from the stored *FCPD* by the *Fetch* method and added to P (Line 6). If *tmp* does not exist in τ_s, then the *Search* method returns $\emptyset$. In this case, all direct subsets of *tmp* are generated by the *Subset* method and pushed into q (Lines 9–10). Finally, P is returned (Line 11).

Algorithm 1. LM Query Processing Algorithm

1: **Input:** Q_{r_0}: a given SRPQ
2: **Output**: P: all frequent co-movement patterns contained in Q_{r_0}
3: $P \leftarrow \emptyset, q \leftarrow Q_{r_0}$
4: **while** $(q \neq \emptyset)$ **do**
5:     $tmp \leftarrow q.pop()$
6:     $p \leftarrow \tau_s.\text{Search}(tmp)$
7:     **if** $p \neq \emptyset$ **then**
8:         $P \leftarrow P \cup \text{Fetch}(F, p.pointer)$
9:     **else**
10:         $q \leftarrow q \cup tmp.Subset()$
11: **return** P

---

5   Temporal Range Pattern Queries (TRPQ)

In this section, we first present the *TFCMP-tree (short for Temporal Frequent Co-Movement Pattern)*. Then, the *Minimum Interval Frequency (MIF)* query processing algorithm is presented to efficiently process *TRPQ* queries using *TFCMP-tree*.

5.1   TFCMP Index Structure

Definition 10 (Minimum Frequency Interval). *Given a timestamp t_i, we define the minimum frequency interval $I_{\alpha,t_i} = [t_i, t_j]$ of a frequent co-movement pattern α such that α is a frequent pattern in any interval $[t_i, t_k]$ if $t_j \geq t_k$, and α is not frequent in any interval $[t'_i, t'_k]$ if $t_j < t_k$.*

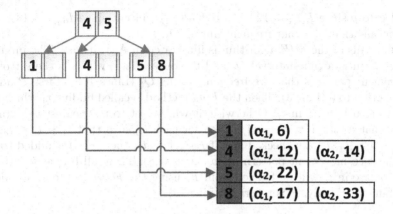

Fig. 4. An instance of the TFCMP-tree

For example, assume $occ(\alpha_1) = ([1,3],[4,6],[8,12],[13,17])$, $occ(\alpha_2) = ([4,11],[5,14],[8,22],[12,33])$, and $\mu = 2$. Then, $I_{\alpha_1,1} = [1,6]$. The *TFCMP-tree* corresponding to this example is illustrated in Fig. 4.

The *TFCMP-tree* τ_t is basically an inverted index on time, where each node $I_{\alpha,T}$ includes all patterns α in $FCPD$ that have an occurrence which starts at time T. For each such pattern α, the pair $(id(\alpha), I_{\alpha,T})$ is stored in $\tau_t[T]$. Moreover, for efficient lookup of the index nodes the nodes are indexed using a B^+-tree with T as the index key. Figure 4 shows an instance of a *TFCMP-tree*.

5.2   MIF Query Processing Algorithm

A naive way to answer time-slice pattern queries is to sequentially read patterns from F and check which ones are frequent during the given time-interval $Q_T = [t_s, t_e]$. However, when F is large, using this approach causes many I/O requests which is time consuming. Also, for each pattern α, $occ(\alpha)$ must be enumerated to see if it is frequent during Q_t or not. To address these issues, we propose the *Minimum Interval Frequency (MIF)* query processing algorithm using the *TFCMP-tree* in order to efficiently retrieve co-movement patterns that are frequent during Q_t.

First, the *MIF* algorithm needs to find $t_1 = t_s$ (or the earliest timestamp after t_s if t_s does not exist in τ_t). In order to speed this up, τ_t contains a B^+-tree that is constructed on timestamps (see Fig. 4). After finding t_1, for each $t_j \in [t_1, t_e]$, the *MIF* algorithm checks the elements of $\tau_t[t_j]$. For each element $(id(\alpha), I_{\alpha,t_j})$, if $I_{id(\alpha),t_j} \le t_e$ then α is frequent in Q_t and $id(\alpha)$ must be added to the result set p. After finding the id of all frequent co-movement patterns in Q_t, actual patterns must be fetched from F. To reduce the fetching time *MIF* uses F', which contains the same set of patterns as F but patterns are sorted based on their ids.

For example, for the given *TFCMP* index τ_t in Fig. 4 if $Q_t = [2,13]$, t_1 will be 4, since $t = 2$ does not exist in τ_t. There are two elements in $\tau_t[4]$. $id(\alpha_1)$ will be added to p since $I_{\alpha_1,4} = 12 \leq 13$ but $id(\alpha_2)$ will not since $I_{\alpha_2,4} = 14 > 13$, which means that α_2 is not frequent during Q_t.

The details of the *MIF* algorithm is illustrated in Algorithm 2. The input to the algorithm is a time-interval Q_t and the output is P which is the set of all co-movement patterns that are frequent during Q_t (Lines 1–2). First P and p are initialized to $\emptyset$ (Line 3). Then the *Find* method is called finding t_1 which will be set to t_s if t_s exist in τ_t. Otherwise, it will be set to the nearest timestamp after t_s that exists in τ_t (Line 4). Then for each t_i, such that $t_1 \leq t_i \leq t_e$, all $(id(\alpha), I_{\alpha,t_i}) \in \tau_t[t_i]$ are checked and if $I_{id(\alpha),t_i} \leq t_e$, then $id(\alpha)$ is added to set of frequent co-movement pattern ids p (Lines 6–8). Then, all frequent patterns whose ids are in p, will be fetched from F' using the *Fetch* method and added to P (Line 9). Finally, P is returned (Line 10).

Algorithm 2. MIF Query Processing Algorithm

1: **Input:** $Q_t = [t_s, t_e]$: a given time-interval
2: **Output:** P: all frequent co-movement patterns during Q_t
3: $P \leftarrow \emptyset, p \leftarrow \emptyset$
4: $t_1 = \tau_t.Find(t_s)$
5: **for** $t_i = t_1 : t_e$ **do**
6:     **for each** $(id(\alpha), I_{\alpha,t_i}) \in \tau_t[t_i]$ **do**
7:         **if** $(I_{\alpha,t_i} \leq t_e)$ **then**
8:             $p \leftarrow p \cup id(\alpha)$
9: $P \leftarrow Fetch(F', p)$
10: **return** P

---

6   Spatiotemporal Pattern Queries (STRPQ)

One approach to answer a spatiotemporal pattern query $Q_{ST} = <Q_r, Q_t>$ is to perform the spacial range pattern query first (as explained in Sect. 4) following by a frequency count for each pattern that meets the spatial criteria to check whether they are frequent during Q_t or not. We propose the *LM-MIF* query processing algorithm which uses both of the two index structures proposed in Sects. 4 and 5 to process spatiotemporal pattern queries.

6.1   LM-MIF Query Processing Algorithm

For a given spatiotemporal query Q_{ST}, first the *LM-MIF* finds all patterns that are completely contained in Q_r using the *LM* algorithm described in Sect. 4. The results of this step is stored in P_1. Next, it finds the id of patterns that are frequent during Q_t using the *MIF* algorithm and stores them in P_2. Then, it returns patterns in P_1 that their ids exist in P_2 and there is no need to perform the frequency count for patterns in P_1. Note that the actual patterns are no longer fetched by *MIF* since they are fetched by *LM*.

7   Experiments

In this section, first we discuss our experimental setup in Sect. 7.1. The results of our empirical performance evaluations of the index construction and query processing with the proposed *FCPIR* framework are provided in Sects. 7.2 and 7.3, respectively.

7.1   Experimental Setup

The *FCPIR* framework is implemented in C# and experiments are carried out on a workstation with Intel Core-i7 3.6 GHz CPU and 16 GB of memory, running Windows 10.

We tested the *FCPIR* framework using the Porto Taxi Dataset [5], which is a real dataset and contains trajectories of 442 taxis traveling in the city of Porto, in Portugal. The dataset was captured during a period of 11 months and includes many features such as latitude and longitude, call type, date, etc.

We used the methods proposed in our prior work [2] to mine all frequent co-movement patterns with $\mu = 15$, $T_{MAX} = 8$, and $n = 30$ from the aforementioned dataset. Then subsets of this *FCPD* dataset with different number of patterns were randomly selected to be used as input for our experiments.

7.2   Index Construction

In this section, we study the performance of the *TFCMP* and *SFCMP* trees by measuring their construction times and their compression ratio to the size of corresponding *FCPD*, as the number of patterns in the dataset grows. The results of this experiment are provided in Figs. 5a and b. In both charts the x-axis shows the number of patterns in the given *FCPD*, while the primary and secondary y-axis in Fig. 5a show the construction time (in milliseconds) for TCMP and TSPQ respectively; finally, the y-axis in Fig. 5b shows the compression ratios of the *TFCMP* and *TSPQ* trees to the size of given *FCFD* in percentage.

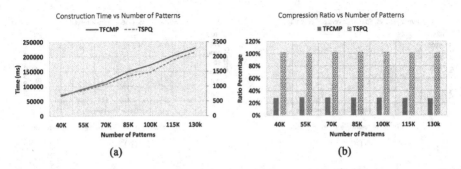

Fig. 5. (a) Construction time in milliseconds and (b) Compression ratio percentage vs. number of patterns in the dataset

As it can be observed in Fig. 5a, the construction times of both index structures grow linearly with the number of patterns in dataset, as expected. The
construction time of the *SFCMP-tree* is smaller than that of the *TFCMP-tree*
up to 10 orders of magnitude, since it reads patterns sequentially from the dataset
and computes their minimum frequency intervals while the *TFCMP-tree* needs
to scan the dataset once for each region sequence in order to find all relevant
patterns; hence more I/O time.

As illustrated in Fig. 5b, the compression ratio remains almost steady for
both trees as the size of the input dataset grows. However, *TFCMP* consumes
less storage as compared to *SFCMP*, since it only stores unique region sequences,
while *SFCMP* stores all occurrences alongside with repetitive pattern ids.

7.3   Query Processing

In this section we discuss the performance of the *LM*, *MIF*, and *LM-MIF* query
processing algorithms in terms of number of required I/O operations to process
queries; note that this metric is proportional to the average query response time.
For all experiments, we used an *FCPD* with 130 K patterns with an average
support of 17 for a pattern.

7.3.1   SRPQ

In this section we compare the performance of the *LM* query processing algorithm with that of the basic approach, where all patterns are scanned and fetched
from the disk in a sequential manner and, those patterns that are infrequent in
Q_r are filtered out. The results for this experiment are illustrated in Fig. 6a,
where the x-axis shows the size of the query Q_r (in percentage of the total size
of the region of interest for the given *FCPD*), while the y-axis shows the average
number of I/O requests made by *LM* (the dashed red line) and its ratio to the
number of I/O requests made by the basic approach (the solid orange line).

As, it can be observed in the figure, the number of I/O requests exponentially
increases with the size of Q_r. This increase is mainly due to the *Subset* method

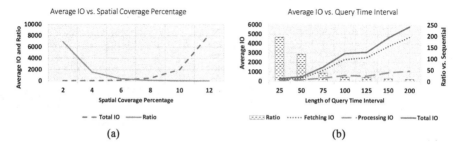

Fig. 6. (a) Average I/O access vs. the length of query time interval, and (b) Average
I/O access vs. the size of spatial range query (Color figure online)

since each generated sub-sequence of the Q_r region sequence must be looked up in the *TFCMP-tree* while processing the query.

7.3.2 TRPQ

For this experiment, we varied the length of query time-interval $|Q_t|$ from 25 to 200 time units (0.5 to 20% of the total temporal length of the trajectories in dataset) and compared the number of I/O requests made by the *MIF* algorithm with the basic approach, where all patterns are scanned and fetched from the disk, and those that are infrequent in Q_t are filtered out. The results of this experiment are illustrated in Fig. 6b, where the x-axis shows $|Q_t|$, while the primary y-axis (on the left) shows the average number of I/O requests made by *MIF*, and the secondary y-axis (on the right) shows ratio of the number I/O request between *MIF* and the basic approach. The solid red line shows the total number of I/O requests made by *MIF*; the breakdown of this cost to the number of I/O requests to find the frequent pattern ids using the *TFCMP-tree* (the dotted blue line), and the number of I/O requests caused by fetching those patterns from the disk (the dashed green line).

As it can be observed, the performance of the *MIF* algorithm is mainly determined by the number of fetch requests, which exponentially increases with $|Q_t|$. The reason is that as $|Q_t|$ increases, the number of patterns that become frequent increases as well; hence, more pages must be fetched from the disk. Moreover, as shown in Fig. 6b, the number of I/O requests grows gradually between $|Q_t| = 100$ and $|Q_t| = 125$. The reason is that in the used *FCPD* the number of frequent patterns for $|Q_t| = 100$ and $|Q_t| = 125$ are not significantly different.

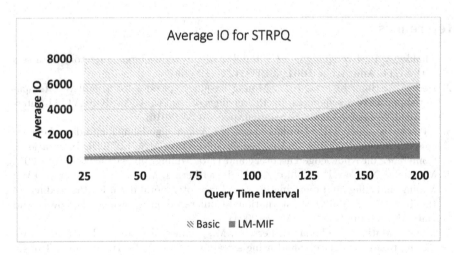

Fig. 7. Average I/O requests for STRPQ vs query time interval (Color figure online)

7.4 STRPQ

In this section we compare the performance of the *LM-MIF* algorithm with the basic approach, where spatial and temporal range patterns queries are processed separately and then their results are merged together. The empirical results for this experiment are illustrated in Fig. 7, where the x-axis shows $|Q_t|$, while the y-axis shows the average number of I/O requests made by the basic approach (the shaded red area) and the *LM-MIF* algorithm (the solid blue area).

As shown in the figure, the number of I/O request made by the *LM-MIF* algorithm (the basic approach) grows gradually (exponentially). The reason is that the *LM-MIF* algorithm does not need to fetch actual patterns from the disk (as opposed to the basic approach) and the pattern ids can be used to filter out infrequent patterns from the result of spatial range pattern query; hence less I/O requests.

8 Conclusion and Future Work

In this paper, for the first time we defined range pattern queries on *frequent co-movement pattern datasets*, and proposed the *FCPIR* framework, including novel index structures and query processing algorithms, for efficient processing of various range pattern queries, namely, spatial range pattern queries, time-slice range pattern queries, and spatiotemporal range pattern queries.

We intend to extend this work in many directions. However, as the very next steps, we will introduce a hybrid tree structure by integrating *SFCMP* and *TFCMP* trees to reduce the storage requirement as well as the number of I/O requests to process spatiotemporal range pattern queries.

References

1. Chakka, V.P., Everspaugh, A.C., Patel, J.M.: Indexing large trajectory data sets with SETI. Ann Arbor **1001**(48109–2122), 12 (2003)
2. Helmi, S., Banaei-Kashani, F.: Mining frequent episodes from multivariate spatiotemporal event sequences. In: Proceedings of the 7th ACM SIGSPATIAL International Workshop on GeoStreaming, p. 9. ACM (2016)
3. Laxman, S., Sastry, P., Unnikrishnan, K.: A fast algorithm for finding frequent episodes in event streams. In: Proceedings of the 13th ACM SIGKDD International Conference on Knowledge Discovery and Data Mining, pp. 410–419. ACM (2007)
4. Mamoulis, N., Cao, H., Kollios, G., Hadjieleftheriou, M., Tao, Y., Cheung, D.W.: Mining, indexing, and querying historical spatiotemporal data. In: Proceedings of the Tenth ACM SIGKDD International Conference on Knowledge Discovery and Data Mining, pp. 236–245. ACM (2004)
5. Moreira-Matias, L., Gama, J., Ferreira, M., Mendes-Moreira, J., Damas, L.: Predicting taxi-passenger demand using streaming data. IEEE Trans. Intell. Transp. Syst. **14**(3), 1393–1402 (2013)
6. Mouza, C., Rigaux, P.: Mobility patterns. GeoInformatica **9**(4), 297–319 (2005)
7. Nanopoulos, A., Manolopoulos, Y.: Efficient similarity search for market basket data. VLDB J. Int. J. Very Large Data Bases **11**(2), 138–152 (2002)

8. Nori, F., Deypir, M., Sadreddini, M.H.: A sliding window based algorithm for frequent closed itemset mining over data streams. J. Syst. Softw. **86**(3), 615–623 (2013)

9. Pfoser, D., Jensen, C.S., Theodoridis, Y., et al.: Novel approaches to the indexing of moving object trajectories. In: Proceedings of VLDB, pp. 395–406 (2000)

10. Sakr, M.A., Güting, R.H.: Group spatiotemporal pattern queries. GeoInformatica **18**(4), 699–746 (2014)

11. Tao, Y., Papadias, D.: Efficient historical R-trees. In: Thirteenth International Conference on Scientific and Statistical Database Management, SSDBM 2001. Proceedings, pp. 223–232. IEEE (2001)

12. Yun, U., Lee, G.: Incremental mining of weighted maximal frequent itemsets from dynamic databases. Expert Syst. Appl. **54**, 304–327 (2016)

Exploratory Spatio-Temporal Queries in Evolving Information

Chiara Francalanci, Barbara Pernici$^{(\boxtimes)}$, and Gabriele Scalia

DEIB, Politecnico di Milano, Piazza Leonardo da Vinci 32, Milano, Italy
{chiara.francalanci,barbara.pernici,gabriele.scalia}@polimi.it

Abstract. Using evolving information within rapid mapping activities in the response phase of emergency situations poses a number of questions related to the quality of information being provided. In this paper, we focus on image extraction from social networks, in particular Twitter, in case of emergencies. In this case issues arise about the temporal and spatial location of images, which can be refined over time as information about the event is being collected and (automatically) analyzed. The paper describes a scenario for rapid mapping in an emergency event and how information quality can evolve over time. A model for managing and analyzing the evolving information is proposed to be used as a basis for analyzing the images quality for mapping purposes.

Keywords: Imprecise spatio-temporal information
Exploratory queries · Evolving information

1 Introduction

Information extracted from social media has proven very useful and informative in many crisis situations [7,18]. In particular, information extracted from Twitter has been studied for its immediacy in making information about the events available.

One of the issues being studied is how to make use of this information within the emergency response activities being activated during an emergency. In particular, we focus on exploiting this information to support rapid mapping activities. Rapid mapping has the goal of providing rescue teams and operators with information about the current situation of the area being interested by the emergency. In the project "Evolution of Emergency Copernicus services" (E^{2}mC^1 the goal is to extend the support the activities of the existing Copernicus Emergency Mapping Service (EMS)^2 providing the rapid mapping operators, who are producing maps of the affected areas based on Earth Observation, with additional information derived from social networks and crowdsourcing, to integrate and complement satellite information available for the service.

---

^{1} https://www.e2mc-project.eu/.
^{2} http://emergency.copernicus.eu/mapping/ems/service-overview.

© Springer International Publishing AG 2018
C. Doulkeridis et al. (Eds.): MATES 2017, LNCS 10731, pp. 138–156, 2018.
https://doi.org/10.1007/978-3-319-73521-4_9

Information derived from social media, and in particular images posted by eyewitnesses, is useful for rapid mapping activities only if associated with an adequate specification of the spatial (geolocalization) and temporal information about the images. However, as we discuss further in this paper, such information may not be available in connection to tweets: only a very small percentage of tweets is geolocalized, the images in the tweets may refer to other areas rather than the geolocation of twitterer, the time and place the image has been taken may not be the same as the time and place of its posting.

In emergency management, often the information related to events becomes clearer only over time, when additional information is collected. For instance, the initial location of the event might be only approximately known initially, and refined over time, or the context of a tweet become known only analyzing a number of associated tweets or other information, which help disambiguating associated information. Also in terms of time information for images, there is a need to distinguish between the time of image posting, the time when the image is taken, the relation to specific events (e.g., in an earthquake tremors sequence, the same building might be in different states after the initial tremor and subsequent tremor events).

The aim of the paper is to discuss the characteristics of information whose spatio-temporal attributes evolve in time. In particular, we analyze how analysis based on imprecise spatial and temporal information can be performed, proposing a meta-model as the basis for the retrieving spatio-temporal information, and discussing how an exploratory approach can support formulating queries and presenting relevant information to operators.

The paper is structured as follows. After discussing relevant state of the art, we present a scenario for illustrating our work in Sect. 3. Then in Sect. 4 we discuss some issues concerning data with imprecise spatial and temporal information. In Sect. 5, we propose a model to support exploratory queries and delineate types of exploration. Finally, we adopt the presented model to discuss two case studies.

2 Related Work

Data quality issues and techniques have been discussed in depth in [4]. Spatio-temporal data quality problem can be found in particular within movement data quality, as discussed in [2], which identifies three quality issues: missing data, accuracy errors, and precision errors. Temporal and spatial resolution, spatial precision, accuracy of positions (such as the accuracy of GPS data or that of the position of a GPS-enabled camera manually set by the operator) affect the quality of information associated to given spatio-temporal coordinates.

Information extraction is the task of extracting structured information starting from unstructured and noisy tweets. For example, in [18] it has been addressed applying conditional random field (CRF), a statistical model which predicts the class of a text token based on its context in the sentence. In general, the named entity recognition (NER) task aims to extract those n-grams which refer to entities of various kinds (people, locations, companies, etc.), and the nature of tweets, very different from traditional texts, poses specific challenges [25]. The recognition task is usually followed by the disambiguation task (named entity linking, NEL) where an n-gram is linked to the exact and unambiguous entry it refers to in an external database.

Focusing on location information, the extraction and disambiguation tasks are also called *geoparsing* and *toponym resolution* respectively. In [14] it has been demonstrated that a state-of-the-art library to perform named entity recognition in a traditional setting (Stanford NER) is not able to deal with bad capitalization, misspellings etc., that are plentiful in microtext. Geoparsing and toponym resolution has been addressed mainly using statistical techniques through trained models [21,23,30], also in combination with heuristics [20]. The social networks have been also used to infer the location of a user [8] or to disambiguate it [15] thanks to the additional contextual information. These researches focus mainly in the user home location rather than the locations mentioned in the messages, but recently social networks analysis has been proven also to overcome the problem of shortness and sparsity of tweet messages analyzing them with respect to other tasks. For example, in [22] interactions and text similarity among tweets are used to improve the topic identification task. Using the social network as additional feature it is possible to refine the extracted information as it grows— that is, as the related event evolves.

Event detection has been performed mainly through clustering [5] and probabilistic models [26], also considering the geographical information [24] and the *spatial density* of the tweets reporting an event [3,28].

Within emergency management, the potential of using data from social media and multiple sources in crisis situation has been studied by several authors [7,17]. The goal in the European E^2mC project is to identify in social media information useful for rapid mapping activities, providing tools that help improving the production times and quality of obtained maps. In [13], we have discussed emergency mapping requirements for building an integrated service-based architecture for E^2mC. Part of the project is focusing on extracting useful information from tweets in the case of earthquakes emergencies with adequate tools. The IMEXT [12] tool environment is a first prototype developed in the project to support Twitter crawling with specific keywords for given event types, geolocating tweets, and extracting images from tweets and from documents linked to the tweets themselves, such as other social media and traditional media.

In recent times, the need is emerging to change the query-answer paradigm common in databases towards exploratory queries, exploratory computing, and exploration systems for big data analysis where the goal is to find interesting patterns or information in large amounts of data (e.g., [9,29]). In their proposal for Queriosity [29], the authors propose a novel approach for developing data exploration systems, to provide insights to users, based on autonomously ranking the relevance of data, learning from users interactions and observation of the environment.

In the present paper, we discuss open issues about the use of information extracted from tweets, with particular reference to their spatial and temporal characteristics, and how an exploratory approach to data can be beneficial for rapidly identifying useful images.

3 Scenario

Copernicus is a European programme aimed to develop European information services based on satellite Earth Observation and in situ (non space) data. Its Rapid Mapping (RM) service provides on-demand and fast provision (within hours or days) of geospatial information as an emergency management service (EMS).

The products provided by the Copernicus EMS Rapid Mapping are standardized with a set of parameters the users can choose requesting them; different products are characterized by different information provided, information quality and time necessary to receive them. In particular, there are three different map types: *Reference* maps, which provide knowledge on the territory using data prior to the disaster and as close as possible to it, *Delineation* maps, which provide an assessment of the event extent (e.g., earthquake impact area map, flooded area map) and optionally its evolution using post-disaster data, and *Grading* maps, which provide an assessment of the damage grade (affecting population and assets like settlements, transport networks, industry and utilities) and optionally its evolution using post-event data. Moreover, maps can be requested in service level 1 (SL1), provided within some hours after delivery and quality approval of imagery, or service level 5 (SL5), provided typically in five working days.

One of the main challenges faced by Copernicus EMS is related to *timeliness*, since it is not unusual to experience delays up to 72 h to receive the first information as satellite information can be incomplete (e.g., due to clouds or delays in receiving information due to satellite passages).

The E^2mC project tries to fill the gap demonstrating the technical and operational feasibility of the integration of social media analysis and crowd-sourced information in the Copernicus EMS improving the timeliness and accuracy of geo-spatial information, particularly in the first hours after the event. Indeed, social media are a relatively new and increasingly important source of information and one of the main advantages is related to timeliness: immediately after an event, large amounts of potentially useful information and media are posted on social media. Processing social media is challenging: it is an example of "big data" with hundreds of millions of posts every day that can be overwhelming and confusing and often include personal impressions rather than useful information. Information on social media is not verified, can be incomplete and partial, even if "some really interesting and important messages do get posted, sometimes providing information that is not available through other channels" [7].

In [12], as a case study, the earthquake in Central Italy of August 2016 has been considered, analyzing the tweets posted just after the event focusing on image extraction of potentially useful and geolocated images. Focusing on image extraction in this context, the goal is to find *useful* images. To consider an image useful, near to an objective usefulness of the image itself, it is necessary to be able to locate it precisely in time and space. However, as detailed in Sect. 4, some important issues are related to the *evolving* and *imprecise* information available on social media, which reflects in the *quality* of the information provided. The recognition and disambiguation phases of the locations cited in the text are prone to imprecisions, so their extraction is an open problem (see Sect. 2). To address the problem, in [27] a new approach has been proposed to improve the recognition and disambiguation performances using both the context provided by the other locations in the same message and the context provided by other messages in the implicit social network related to each message. In the following the locations will be extracted from text using this algorithm.

4   Evolving and Imprecise Information

In this paper we focus on representing and analyzing evolving information, and specifically on spatio-temporal aspects of information. We define as spatio-temporal evolution of information the process of refining the information on a specific event over time. Information becomes more precise as the event evolves, e.g., the location of the event becomes more precise over time, delineating the area of interest, tweets can be geolocated using context information provided from other sources or extending them with information derived from their analysis, external additional information may become available.

4.1   Imprecise Information

As mentioned in Sect. 2, several authors have proposed approaches to exploit tweets as information sources during emergencies and there is a wide literature about quality of spatial and temporal information.

Even if theoretically an event is defined as something which happens at a precise time and in a delimited space (for instance, the EMSR177 activation for the Central Italy earthquake is associated with an Event Time (UTC): 2016-08-24 01:36, it has an Area Descriptor: Lazio, Abruzzo and Umbria Regions, and it is associated to an Activation Extent Map that provides the polygons for delimiting the areas of the grading maps with their geographical coordinates), the information extracted from social media and the nature of real events themselves bring a series of *imprecisions* with respect to the availability of both spatial and temporal information, which is necessary to take into account and address performing rapid mapping.

Spatial Information. Geographical information is a rare resource on social media. Only 0.5%–2% of tweets are geotagged [7,20], and the metadata associated to images cannot be used since images loose all metadata, including their geographical coordinates, when stored by Twitter.

Therefore, *geolocation*, that is the activity of associating a location to the messages using other indicators like the text, the social networks, the URLs contained in the message, etc., becomes crucial. In particular, it has been demonstrated the value of extracting the locations referenced in the text [20,30], firstly *recognizing* the toponyms mentioned and then *disambiguating* them to the exact locations they refer to. However, this task introduces imprecisions due to ambiguities which exist among location names and common names (*geo/non-geo* ambiguities) and among location names themselves (*geo/geo* ambiguities). Extracting locations from tweets has additional challenges with respect to traditional texts because the short, noisy and decontextualized nature of tweets.

Even if the locations mentioned in tweets are correctly recognized and disambiguated, they are typically imprecise, at some extent, for rapid mapping purposes. Indeed, while the coordinates associated to a geotag attached by Twitter should precisely identify the location where the tweet has been submitted, the locations mentioned in the text could be more general, citing for example a city or a region or could contain multiple references to locations, also with different levels of granularity.

A related challenge comes from the *gazetteer* used, which could not cover equally all the target locations and could contain errors. Indeed, "the output of any geocoding algorithm is only as exact as the knowledge base that underlies it" [30]. For example, GeoNames does not contain many street/road names or point of interests. Moreover, a gazetteer like GeoNames contain only few "alternative names" for each entry and they do not account for any possible way each location could be referred to. More precise information for geolocating tweets is provided by OpenStreetMap[3], however the richness of its information provides additional more precision, but also additional disambiguation problems, as the geographical names can be found also in streets and points of interest. For the analysis in this paper, locations are derived using an approach based proposed in [27], based on Named Entity Recognition libraries [1] and GeoNames (in the first case study) and OpenStreetMap (in the second case study) as gazetteers.

Another imprecision is related to the location of the events themselves. Indeed, it is not always trivial to define precisely the boundaries of an event or establish whether two near events are actually the same one. For example, considering an earthquake, the most affected areas typically are those nearer to the epicenter. However, the actual damages will depend also from other factors, like the state of the affected buildings and the population density, therefore the most affected areas could not be spatially continuous. An earthquake is typically felt also at many kilometers of distance, causing only minor damages in more distant areas, therefore is not easy to establish the boundaries of the event to monitor. An example is given by the earthquake of 24th August in Italy. The epicenter is in Accumoli, a small municipality of 650 inhabitants, but the shake has been felt distinctly also in other near areas, in particular Rome, even if no significant damages were reported there. However, the fact that Rome is the most populated city of Italy, brought a significant number of reports from and

---

[3] https://www.openstreetmap.org.

about that city, so that, especially in the first hours, simply monitoring Twitter it seems like the main target of the earthquake is indeed Rome instead of Accumoli.

Temporal Information. It is not trivial to precisely assess the duration of an event in terms of its starting and ending time. If a sudden event like an earthquake has at least a clear starting point, other events like floods could start slowly as simple rains. Moreover, an event could be characterized by several subevents [7]: for example several aftershocks of an earthquake could be separated even by hours or days. Consider for example the sequence of shakes related to the earthquake of August, Italy, shown in Fig. 1. It is challenging to precisely distinguish the subevents starting from tweets: for example, after a shake, the posts and the related images could be related to a previous shake, and, even if apparently useful, being indeed outdated as information.

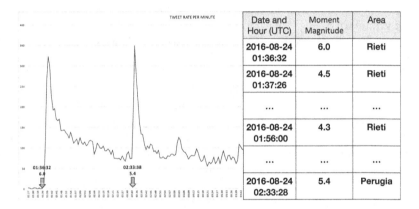

Date and Hour (UTC)	Moment Magnitude	Area
2016-08-24 01:36:32	6.0	Rieti
2016-08-24 01:37:26	4.5	Rieti
...	...	...
2016-08-24 01:56:00	4.3	Rieti
...	...	...
2016-08-24 02:33:28	5.4	Perugia

Fig. 1. Time of events [16]

Therefore, assigning a precise time to information or media items could be challenging and often the only known information is an upper bound.

Following the temporal database concepts described [10], it is also necessary to distinguish the time of occurrence of events or of the information about the real world from the time in which they are recorded in the system (transaction time). In general, an evolution in time is associated to all information to be analyzed, in particular concerning the precision and accuracy of available information.

5   A Model for Spatio-Temporal Imprecise Information

In this section, we propose a model for spatio-temporal information related to events that can be used to support queries and exploratory queries on an event. First, we describe the model in Sect. 5.1, then we propose some directions for exploring the available data.

5.1 Modeling Spatio-Temporal Information

The goal of the model is to support information related to events that can become more precise over time and to be able to associated to it time and space information. The model, illustrated in Fig. 2, is centered on the concept of *event*, which may have subevents. We assume an occurrence time and location for the event are available, as a starting point for the exploration. To each event, related *documents* are associated, composed of *items* (carrying information) which are progressively located in time and space, as related information (*info*) becomes available (or may be automatically derived). The time of the document production is also recorded. This is not necessarily the time to be associated to its items (e.g., an image included in a tweet could have been produced at an earlier time). The documents have an *author*, whose location in time and space may be also known with different degrees of precision and could be also be variable in time. It has to be noted also that the location associated to the profile of the author could be available, but could be also misleading in some cases if it is the home location and not the posting location. Time and location associations are characterized by two attributes, *precision* and *accuracy* as it is customary in the data quality literature (see Sect. 2).

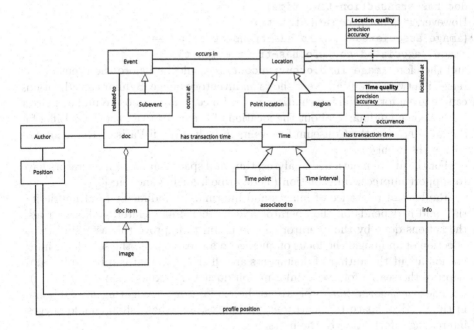

Fig. 2. Event exploration model

5.2   Exploratory Queries

The risk of information overload and, on the other hand, of having too little information is present. For instance, in the first hours after the event, the contribution in the social media of new images (i.e., not duplicates of previous ones) and related to the event might be limited.

Initially, the exploration of available data (which can be massive, see for instance the tweet rates for the event in the scenario, shown in Fig. 1) has the goal of delineating the affected areas in order to facilitate rapid mapping preparation.

Immediately afterwards, the goal of the mapping operator is to find useful images. Sometimes one (or very few) image is enough (e.g., to confirm the forecast of a flood, showing initial damages, or if an image shows clearly an affected area). In other cases, the goal is to identify twitters who are eyewitnesses and therefore giving direct information on the event. The temporal and spatial information associated to documents, document items, and authors can all be useful to support the exploration. When looking for images, the goal is to identify images which are produced after the event, i.e.,

(`image.associated-to.info.occurrence.time`>`event.occurs-at.time`).

The time of an image precedes the time of the document in which it is inserted

(`image.associated-to.info.occurrence.time`<

`doc.has-transaction-time.time`).

However, it is not necessarily true that

(`image.associated-to.info.has-transaction.time`=

`image.associated-to.info.has-transaction.time`)

and therefore `image.info.occurrence.time` could be before the event occurrence, if a tweet about an event shows an inventory image. Similar considerations can be given for spatial information: for instance, it is common to find as twitter images some landmarks about the location of the event (e.g., the Tour Eiffel for events in France, the Coliseum for events in Italy, even if Paris or Rome are not within the event area).

For the above mentioned analyses, time and space comparison operators have to support imprecise information (as described, for instance, in [6]).

Ranking of relevance of images and information is difficult (and multicriteria), and it depends on the operator and his/her area of interest, however also the actions done by the operator can be useful to improve this assessment: the selection of an image, the focus of queries on a specific area, the search for information about the author of documents are all elements that can be exploited to improve the search for useful information during the exploration.

The *query formation* process can be exploratory, navigating available info in time and in space to identify useful images, and the system should suggest interesting information to the users.

Type of queries include:

– Spatial queries to delimit area of interest.
– Event time and space derivation (through constraints from document information).
– Evolution in time of a point /area in space.

Information suggestions, not related to a user-generated query, but autonomously provided by the system, might include selection of images by relevance to be proposed to operators, delineation of areas, identification of Twitter influencers eyewitnesses for the event, and so on.

6 Case Studies

In this section, we discuss some aspects related to the use of the previously described model to explore information about emergency events. We consider case studies from three activation of the EMS rapid mapping service of Copernicus, two related to the 2016 Central Italy Earthquake, and the third one to the 2014 floods in Southern England.

6.1 Central Italy Earthquake

In this section, we discuss the evolution of tweets in two EMS activations for the earthquake in Central Italy. After the first activation EMSR177 described in Sect. 3, another activation was started in October (EMSR190), following a major aftershock on October 26.

Event 1 - August 24, 2016. In this event, using the crawling techniques illustrated in [12], about 150,000 possibly relevant tweets have been extracted from the 48 h immediately following the event.

The amount of tweets (*documents* in the model described in Sect. 5) with locations recognized and disambiguated from text (using GeoNames as gazetteer) increases over time, considering the additional context provided by new tweets posted (see [27]). For example, focusing on the tweets posted in the first 30 min after the earthquake, 28.1% of them have locations recognized and disambiguated thanks to the context they provide to each other. 48 h later, among *the same tweets* posted in the first 30 min, additional 3% of tweets have locations recognized and disambiguated thanks to the additional context. As comparison, only 0.35% of tweets are geotagged in this dataset.

In addition to the locations associated to individual tweets, it is interesting to evaluate the location related to the event itself. As mentioned in Sect. 4, considering the cumulative number of the geotagged tweets only, the location related to the epicenter, Accumoli, is not highlighted with respect to, for example, Rome, which is a not damaged but much more populated location. If, instead, the locations referenced in the text are used (extracted as explained in [27]), the situation improves, as shown in the graph in Fig. 3. Indeed, there is an improvement in terms of the *volume* of locations, which are significantly (orders of magnitude) more, and in the *rate of growth* of the reports related to Accumoli with respect to those related to Rome, so that after about 3 h Accumoli is highlighted as main location target of the event. It is interesting to notice that, as the time passes, the cumulative number of reports related to Accumoli become greater and greater, making the situation more and more clear. However, there exists a

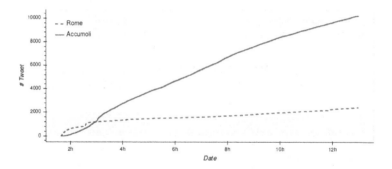

Fig. 3. Cumulative number of tweets about Rome and about Accumoli according to the text-extracted locations.

"window of uncertainty" in the first three hours where Accumoli does not emerge yet as main location.

The first tweets carrying useful information with images started to arrive early, even if the event occurred at night (see for instance Fig. 4). Several tweets contained the photograph shown in the figure, which was taken about two hours after the event. It has to be noted that the image was therefore available many hours before the EMS activation, which officially started more than 8 h after the event).

Fig. 4. Image extracted from tweets in the first hours after the event of August, 24.

It is interesting to analyze the tweets associated to the image in Fig. 4, to show the related challenges and their contribution to address time and space imprecisions in the proposed model. They are shown, along with their timestamp and the profile location of the authors, in Table 1[4].

Analyzing them, several considerations arise. First of all, no tweet is geotagged, so that the positions of the authors submitting the tweets is unknown. Except tweets #3 and #4, they all have different authors. The locations associated to the authors profiles are not useful: they are missing or not real locations (MotoriNews24 & KeepRadio) or too coarse-grained (low precision, e.g., the province) and in any case incoherent. The tweet #1 is posted about two hours after the earthquake, describing the location generically as "the road to Norcia". It is a very imprecise information: there exist many roads to Norcia each one traveling many kilometers. Therefore, one could simply take Norcia itself as location for the image—even if it is more than $275\,\mathrm{km}^2$—accepting a low precision and without being able to quantify the accuracy. Regarding the time associated to the image, this tweet gives an "upper bound" to its creation, so that the associated time is an interval of about two hours with high precision[5]. The tweet #2 cites Norcia again so it does not add significant information regarding the location or the time of the image. The tweet #3 is particularly interesting because it is posted just after half an hour the first tweet, but it better describes the location. Indeed, it cites the precise road ("Salaria SS4"), which is a road of more than $200\,\mathrm{km}$, citing exactly the point ("at Sigillo"). There are several locations named "Sigillo" in Italy: the tweet cites other locations, and in particular "Posta", indicating that it refers to Sigillo part of the municipality of Posta, which has just 151 inhabitants, and not, for example, Sigillo municipality in the province of Perugia, which is more populated (population 2468). The approach described in [27] (see Sect. 3) is able to correctly recognize and disambiguate Sigillo, thanks to a spatial correlation found among the different locations mentioned in the tweet. The road SS4 at Sigillo is about 5 km, so using this tweet it is possible to associate a much better precision to the image, even if the accuracy is still unknown. This tweet does not add precision to the time associated to the image since it is posted later the first one. The following tweets do not add precision since they do not add details regarding the space or time of the image.

Concerning the images associated to the tweets, as illustrated in [12], only about 19% of the images extracted from geolocated tweets have been considered potentially useful, i.e., showing damages. As a document component we also considered videos. The analysis of videos showed that, among the around 1200 videos, very few of them were interesting for mapping purposes, as most of them were showing interiors. However, it has to be noted that some of these videos

---

[4] Notice that they are not the *complete* set of posted tweets with that image. Some tweets could not be retrieved and $\approx 2.7\%$ of the retrieved tweets are not available anymore. Moreover, to find also the same image at different resolutions or with slight modifications a hashing algorithm has been used and it has false negatives.

[5] This is true if it would be possible to confirm that the image comes from the target event; in general the starting point of the interval is unknown or, equivalently, the precision associated to the interval of two hours is low.

Table 1. All the retrieved tweets associated to the image in Fig. 4.

#	Time (UTC+2)	Text	Profile location
1	5:41:37	#terremoto Crolli lungo strada per Norcia. Mandateci le vostre foto dall'Umbria a redazione@umbria24.it o sui social	Perugia
2	5:43:32	Strada per #Norcia crollo #terremoto	spello
3	6:11:57	Questa la Salaria SS4 all'altezza di Sigillo ... #terremoto #amatrice #Accumoli #Posta #ArquatadelTronto	MotoriNews24 & KeepRadio
4	6:37:57	RIPETO!!! FONDAMENTALE LASCIARE LIBERA LA SALARIA!!! #terremoto SE NON NECESSARIO NON USATE LA SALARIA SS4 !!!	MotoriNews24 & KeepRadio
5	6:45:26	Terremoto ecco la salaria all'altezza di sigillo #terremoto @SkyTG24 @RaiNews	
6	7:17:34	?????#Terremoto #Reatino invitiamo a non recarsi nelle aree terremotate per agevolare soccorsi #InfoAstral	Lazio Italia
7	7:25:05	La galleria sulla Salaria per Ascoli all'altezza di Sigillo #terremoto	

were very informative, showing areas from aerial images (mainly from drones) from official sources which reported also the exact location.

This analysis reinforces the need to support an exploratory search for data: the goal is not of retrieving a large number of images or videos, but of identifying the areas for which such images are needed for mapping, and in selecting those images which are more informative for the operators of the mapping activity. An analysis is ongoing for supporting the relevance ranking with statistical and context data: number of retweets, type of source (official sources, local sources), finding them through exploratory methods supported by automatic analysis of the sources and by following operators decisions (selection of an image for mapping activities, focus on a given area, and so on).

Event 2 - October 26, 2017. A second activation for the same area followed the first one, after a major aftershock. From this activation we have derived a negative result: only a very small (20) number of geolocated images where extracted, many of them not useful. The reasons for this result are probably due to the fact that at that moment the interested areas had been evacuated because of damages to buildings. It has to be noted also that even if another major aftershock was recorded on October 30, no new EMS activation was started. This fact supports the fact that in this special case the need for rapid mapping in already damaged areas becomes less urgent, while other types of maps need to be produced (e.g., assessment maps which are produced after the events), which are not the subject of this study.

6.2 South England Floods

As a second case study, we present an example for the use of exploratory queries in space and time in a case of a rapid mapping activation in a flood emergency, in particular EMS activation MSR069: Floods in Southern England.

In the case of rapid mapping for floods, the goal is to produce delineation maps indicating the area which has been affected by the floods. The event started on Feb. 10, 2014 and 22 delineation maps were produced in the period Feb. 12 to Feb. 25. The rectangles on the map of Fig. 5 show the areas for which maps were produced.

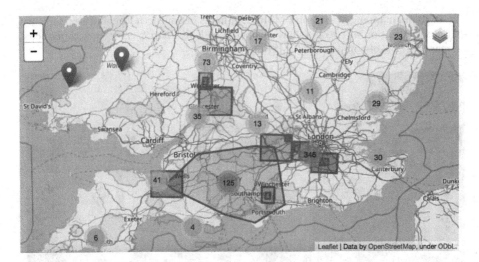

Fig. 5. Exploring tweets related to the EMSR069 activation - spatial exploration.

In the following, we focus our exploration of the area at the east of London, including the towns of Datchet and Egham (indicated in EMS as Steines area in delineation maps). A close up view of the detailed area is shown in the delineation map of Fig. 6.

As reported from the rapid mapping experts in [11], the creation of maps from satellite data poses different problems in floods from the case of earthquakes, where damage can often be seen clearly from satellite views, as in this case the goal is to detect water which is anomalous, e.g. on streets and roads, and the analysis of satellite images is not always sufficient to detect flooded areas in urban contexts.

As example, in the maps shown in Fig. 6, flooded areas are mainly located in fields surrounding the built environment. One of the challenges of the E2mC

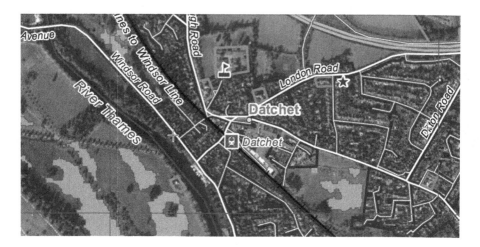

Fig. 6. Rapid mapping activation EMSR069: Floods in Southern England - exploratory spatial visualization of tweet (source Copernicus EMS http://emergency.copernicus. eu/mapping/list-of-components/EMSR069).

Fig. 7. Precisely located flooded road image from Tweet: Queens Road, Datchet CP, Datchet, Windsor and Maidenhead, South East, England.

project is to leverage social media to provide additional information about flooding of roads and streets. An example of image which is considered useful in this context as an additional source is shown in Fig. 7, with a flooded street. However, this image is useful for the rapid mapping operators only if the image is precisely geolocated. However, from the 108,757 tweets we examined from the Southern England case study, only 3,333 are geotagged (310 geotagged images), and from

the analysis of their characteristics most of them are not useful since they are
not photos or the images are not useful. In this case we refined our geolocating algorithm using NER, and in particular Stanford CoreNLP (Named Entity
Recognizer module) for extracting entity names which are then recognized using
Openstreet map as a gazetteer, since it provides a more precise location to the
street level. To minimize false positives, we chose to examine only the tweets
which had two related named entities in the text or hashtags. The resulting
tweets containing images for the area of interest are 378. The images have then
be examined by the crowd to eliminate clearly useless images for the area which
has been studied in detail.

In the following, we show two applications of the exploratory mechanisms
as discussed in the previous section of the paper. At this moment, we create
separately two types of exploration maps, one for temporal exploration and a
second one for spatial exploration. In the future implementations we are planning
to integrate the exploration mechanisms.

One of the characteristics of information derived from social media, is that
the number of potentially useful information is very variable in space and time.
Therefore, first, we let the user explore the data temporally. In this case, the
tweets can clearly show the evolution of tweets related to the flood. For instance,
using an incremental exploration space of three hours, we can see that almost
no geolocated tweets with images were available in the first hours of the event
(during the night), and that the incremental exploration can show the evolution
of the flood. For instance, we can easily identify a surge of tweets in the morning
of Feb. 11, as shown in Fig. 8.

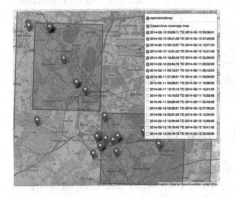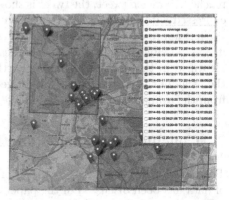

Fig. 8. Tweets related to the EMSR069 activation in the Steines area - temporal exploration.

Concerning the spatial exploration, in Fig. 9, the number of tweets to be
explored is shown in the dots. To allocate the dots on the map we have used
the "marker clustering" functionality[6] for Leaflet[7]. When passing on a dot, the
polygon of the area of the tweets numbered in the dot is shown.

---

[6] https://github.com/Leaflet/Leaflet.markercluster.
[7] http://leafletjs.com/.

Fig. 9. Exploring tweets related to the EMSR069 activation in the Steines area - spatial dimension.

Selecting a detailed view, such as the Steines area shown in Fig. 9, more details are shown on the specific area, until the single tweets (shown as flags) are reached. For each tweet, the tweet is shown together with its derived geolocation, expressed with a geohierarchy, as shown in the caption of Fig. 7.

The approach proved useful in its first experimentation on this case. In fact, it was possible to identify flooded areas in an urban setting which were not identified in the delineation maps. Further work is needed in order to improve the recall of the current algorithm for geolocation, to gather further information from the available tweets on the emergency event.

7 Concluding Remarks and Future Work

The need for facilitating the retrieval of useful information from large amounts of data has been discussed in the paper, focusing on spatial and temporal information extracted from Tweets. As discussed in the paper, while there is a wide margin for improvement of the automatic geolocation of information, the focus of the discussion is about exploiting already available information, which might evolve over time, applying an exploratory approach to data analysis to find useful information for the tasks to be performed.

Acknowledgments. This work has been partially funded by the European Commission H2020 project E^2mC "Evolution of Emergency Copernicus services" under project No. 730082. This work expresses the opinions of the authors and not necessarily those of the European Commission. The European Commission is not liable for any use that may be made of the information contained in this work. The authors thank Paolo Ravanelli for his support in creating event-specific crawlers, and Paolo Gugliemino and Matteo Montalcini for their work on the case study.

References

1. Al-Rfou, R., Kulkarni, V., Perozzi, B., Skiena, S.: Polyglot-NER: massive multi-lingual named entity recognition. In: Proceedings of the 2015 SIAM International Conference on Data Mining, Vancouver, British Columbia, Canada, 30 April – 2 May 2015, April 2015
2. Andrienko, G.L., Andrienko, N.V., Fuchs, G.: Understanding movement data quality. J. Locat. Based Serv. **10**(1), 31–46 (2016). https://doi.org/10.1080/17489725.2016.1169322
3. Andrienko, N., Andrienko, G., Fuchs, G., Rinzivillo, S., Betz, H.D.: Detection, tracking, and visualization of spatial event clusters for real time monitoring. In: IEEE International Conference on Data Science and Advanced Analytics (DSAA), 2015. 36678 2015, pp. 1–10. IEEE (2015)
4. Batini, C., Scannapieco, M.: Data and Information Quality - Dimensions, Principles and Techniques. DSA. Springer, Cham (2016). https://doi.org/10.1007/978-3-319-24106-7
5. Becker, H., Naaman, M., Gravano, L.: Beyond trending topics: real-world event identification on Twitter. In: ICWSM, vol. 11, pp. 438–441 (2011)
6. Brusoni, V., Console, L., Terenziani, P., Pernici, B.: Qualitative and quantitative temporal constraints and relational databases: theory, architecture, and applications. IEEE Trans. Knowl. Data Eng. **11**(6), 948–968 (1999). https://doi.org/10.1109/69.824613
7. Castillo, C.: Big crisis data: social media in disasters and time-critical situations (2016)
8. Davis Jr., C.A., Pappa, G.L., de Oliveira, D.R.R., de L Arcanjo, F.: Inferring the location of Twitter messages based on user relationships. Trans. GIS **15**(6), 735–751 (2011)
9. Di Blas, N., Mazuran, M., Paolini, P., Quintarelli, E., Tanca, L.: Exploratory computing: a comprehensive approach to data sensemaking. Int. J. Data Sci. Anal. **3**(1), 61–77 (2017). https://doi.org/10.1007/s41060-016-0039-5
10. Dyreson, C., Grandi, F., Käfer, W., Kline, N., Lorentzos, N., Mitsopoulos, Y., Montanari, A., Nonen, D., Peressi, E., Pernici, B., et al.: A consensus glossary of temporal database concepts. ACM Sigmod Rec. **23**(1), 52–64 (1994)
11. E2mC Team: Analysis of Copernicus Witness integration issues, E2mC deliverable D1.3, April 2017
12. Francalanci, C., Guglielmino, P., Montalcini, M., Scalia, G., Pernici, B.: Imext: a method and system to extract geolocated images from tweets analysis of a case study. In: Research Challenges in Information Science (RCIS), 2017 IEEE Eleventh International Conference on Research Challenges in Information Science, Brighton, UK, May 2017. IEEE (2017)
13. Francalanci, C., Pernici, B.: Data integration and quality requirements in emergency services. In: Advances in Service-Oriented and Cloud Computing. Springer (in press)
14. Gelernter, J., Mushegian, N.: Geo-parsing messages from microtext. Trans. GIS **15**(6), 753–773 (2011)

15. Ghufran, M., Quercini, G., Bennacer, N.: Toponym disambiguation in online social network profiles. In: Proceedings of the 23rd SIGSPATIAL International Conference on Advances in Geographic Information Systems, p. 6. ACM (2015)
16. Guglielmino, P., Montalcini, M.: Extracting relevant content from social media for emergency management contexts. Master thesis, Politecnico di Milano (2016)
17. Imran, M., Castillo, C., Diaz, F., Vieweg, S.: Processing social media messages in mass emergency: a survey. ACM Comput. Surv. (CSUR) **47**(4), 67 (2015)
18. Imran, M., Elbassuoni, S.M., Castillo, C., Diaz, F., Meier, P.: Extracting information nuggets from disaster-related messages in social media. In: Proceedings of ISCRAM, Baden-Baden, Germany (2013)
19. Imran, M., Mitra, P., Castillo, C.: Twitter as a lifeline: human-annotated Twitter corpora for NLP of crisis-related messages. arXiv preprint arXiv:1605.05894 (2016)
20. Inkpen, D., Liu, J., Farzindar, A., Kazemi, F., Ghazi, D.: Detecting and disambiguating locations mentioned in Twitter messages. In: Gelbukh, A. (ed.) CICLing 2015. LNCS, vol. 9042, pp. 321–332. Springer, Cham (2015). https://doi.org/10.1007/978-3-319-18117-2_24
21. Liu, F., Vasardani, M., Baldwin, T.: Automatic identification of locative expressions from social media text: a comparative analysis. In: Proceedings of the 4th International Workshop on Location and the Web, pp. 9–16. ACM (2014)
22. Nugroho, R., Yang, J., Zhao, W., Paris, C., Nepal, S.: What and with whom? Identifying topics in Twitter through both interactions and text. IEEE Trans. Serv. Comput. **PP**(99), 1 (2017)
23. Paradesi, S.M.: Geotagging tweets using their content. In: FLAIRS Conference (2011)
24. Reuter, T., Cimiano, P.: Event-based classification of social media streams. In: Proceedings of the 2nd ACM International Conference on Multimedia Retrieval, p. 22. ACM (2012)
25. Ritter, A., Clark, S., Etzioni, O., et al.: Named entity recognition in tweets: an experimental study. In: Proceedings of the Conference on Empirical Methods in Natural Language Processing, pp. 1524–1534. Association for Computational Linguistics (2011)
26. Sakaki, T., Okazaki, M., Matsuo, Y.: Tweet analysis for real-time event detection and earthquake reporting system development. IEEE Trans. Knowl. Data Eng. **25**(4), 919–931 (2013). https://doi.org/10.1109/TKDE.2012.29
27. Scalia, G.: Network-based content geolocation on social media for emergency management. Master thesis, Politecnico di Milano, April 2017
28. Tamura, K., Ichimura, T.: Density-based spatiotemporal clustering algorithm for extracting bursty areas from georeferenced documents. In: 2013 IEEE International Conference on Systems, Man, and Cybernetics (SMC), pp. 2079–2084. IEEE (2013)
29. Wasay, A., Athanassoulis, M., Idreos, S.: Queriosity: automated data exploration. In: Carminati, B., Khan, L. (eds.) 2015 IEEE International Congress on Big Data, New York City, NY, USA, 27 June – 2 July 2015, pp. 716–719. IEEE (2015). https://doi.org/10.1109/BigDataCongress.2015.116
30. Zhang, W., Gelernter, J.: Geocoding location expressions in Twitter messages: a preference learning method. J. Spat. Inf. Sci. **2014**(9), 37–70 (2014)

Efficient Cross-Modal Retrieval Using Social Tag Information Towards Mobile Applications

Jianfeng He[1], Shuhui Wang[1(✉)], Qiang Qu[2], Weigang Zhang[3], and Qingming Huang[1]

[1] Key Lab of Intelligent Information Processing of Chinese Academy of Sciences (CAS), Institute of Computing Technology, CAS, Beijing 100190, China
jianfeng.he@vipl.ict.ac.cn, wangshuhui@ict.ac.cn, qmhuang@ucas.ac.cn
[2] The Global Center for Big Mobile Intelligence, Frontier Science and Technology Research Centre, Shenzhen Institutes of Advanced Technology, CAS, Shenzhen 518055, China
qiang@siat.ac.cn
[3] School of Computer Science and Technology, Harbin Institute of Technology, No. 2 West Wenhua Road, Weihai 26209, China
wgzhang@hit.edu.cn

Abstract. With the prevalence of mobile devices, millions of multimedia data represented as a combination of visual, aural and textual modalities, is produced every second. To facilitate better information retrieval on mobile devices, it becomes imperative to develop efficient models to retrieve heterogeneous content modalities using a specific query input, e.g., text-to-image or image-to-text retrieval. Unfortunately, previous works address the problem without considering the hardware constraints of the mobile devices. In this paper, we propose a novel method named Trigonal Partial Least Squares (TPLS) for the task of cross-modal retrieval on mobile devices. Specifically, TPLS works under the hardware constrains of mobile devices, i.e., limited memory size and no GPU acceleration. To take advantage of users' tags for model training, we take the label information provided by the users as the third modality. Then, any two modalities of texts, images and labels are used to build a Kernel PLS model. As a result, TPLS is a joint model of three Kernel PLS models, and a constraint to narrow the distance between label spaces of images and texts is proposed. To efficiently learn the model, we use stochastic parallel gradient descent (SGD) to accelerate the learning speed with reduced memory consumption. To show the effectiveness of TPLS, the experiments are conducted on popular cross-modal retrieval benchmark datasets, and competitive results have been obtained.

Keywords: Cross-modal retrieval · Multimedia
Partial least squares · Images and documents

C. Doulkeridis et al. (Eds.): MATES 2017, LNCS 10731, pp. 157–176, 2018.
https://doi.org/10.1007/978-3-319-73521-4_10

1   Introduction

With rapid advance in Internet technology and data device, amount of data is soaring exponentially. Among this, mobile data is one of main data source created by people. For instance, it is at least 5 hundred million images that are uploaded to the Internet everyday; it is around 20-h videos that are shared in each minute. Furthermore, mobile data tends to appear in the form of multimedia, such as texts, images, sounds and videos which are often applied to record the users' mood in Facebook, Twitter and other social application. It is also mentionable that the mobile data is often illustrated in two or more multimedia. Facebook and Twitter, meanwhile, are also strong evidence to support this point. A microblog of the Facebook is always finished through texts and images, and a video of the Twitter tends to include video, sound and even text. Thus, according to the current situation that heterogeneous modal data is used to describe a theme or one thing, the traditional content-based retrieval in single modality may not fulfill the users' requirement, and it is crucial and imperative to achieve cross-modal retrieval in mobile devices for users. Cross-modal retrieval is a newly proposed retrieval to use one modal query to retrieval the other one modal data [3,4,8,12,30,33]. To be specific, given a text query, then return content-related images; or given a music query, then return content-related video. An concrete example of cross-modal retrieval shown in the Fig. 1. However, the text features and the image features are in different feature space so that they can not be matched directly with each other. Hence, the key problem for cross-modal retrieval is achieving consistent feature representation for each heterogeneous modality [10,13–15,23,26–29,31,35].

Besides the above key problem, characteristics of mobile device, however, also bring more limitation when we apply the application of the model to the mobile devices. There are two main special points in mobile device: small memory (random access memory) and no GPU, thus it is unpractical to achieve the cross-modal retrieval by deep learning frameworks in mobile devices. The first contributing factor is that the deep learning asks for large memory which is at least several hundred megabytes to store its neural weights. The second contributing factor is that the GPU which does not exist in mobile devices is necessary to deep learning to accelerate calculating speed sharply. Then, though the weights of convolutional neural networks (CNN) can be stored in some high-performance mobile devices, it is slow for them to get the CNN features without the GPUs in the testing period. As a consequence of these two points, low hardware condition should also be taken into consideration when we design a model to solve cross-modal retrieval on mobile device. Specially, we put the training process of our model on a computer, and we care the efficiency and feasibility of testing process which has been done on the android virtual device (AVD) of Android Studio.

To learn the consistent feature representation based on low hardware condition, we just give up the deep learning framework but choose traditional subspace learning [1,2,5,9,10,15,22,32–34], which has been a common method to solve cross-modal retrieval and requires low hardware condition. To clarify our idea

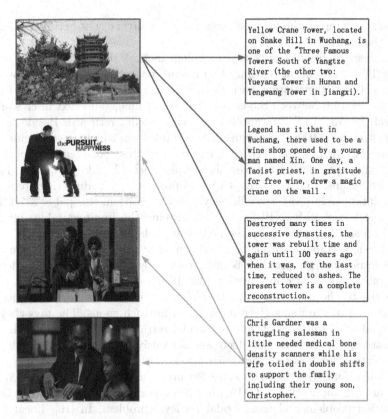

Fig. 1. The examples of cross-modal retrieval between the image and text. The orange rectangles, having common semantics in the Tower of Yellow Crane, show the image-to-text process: given an image, then return the related texts. As for the blue rectangles, which possess semantics in the movie, they illustrate the text-to-image process: return the related images according to the text. (Color figure online)

clearly, we use image-text retrieval as an example to describe our model, but the model is also suitable to other heterogeneous modalities.

The subspace learning methods learn a common feature space so as to match the image and text feature directly and preserve the correlations between image-text pairs. As one of the subspace learning methods, canonical correlation analysis (CCA) [7,15,19] projects the two modal features to a shared latent space which maximizes the correlations between two modalities. So far, many extensions of CCA have been used on similar area. In [15], the semantic correlation match (SCM) was proposed to get a semantic subspace by using logistic regressor based on CCA. In Verme [23] work, correlated semantic representation (CSR) was proposed to obtain a joint image-text representation and an unified formulation by learning a compatible function based on structural support vector machine (SVM). Besides, biliners model (BLM) [22] is also a kind of traditional subspace learning methods, which gets rid of the diversity of between the different modal features.

Another traditional subspace learning method is partial least squares (PLS) [16,17] aiming at learning two latent spaces by maximizing the correlations between their latent variables. Sharma *et al.* [19] applied the PLS to build the relation between the latent variables of image and text in cross-media retrieval. In [10], PLS was applied into cross-media retrieval. Besides solving the cross-modal retrieval problem, PLS also has been widely used in multi-view problem [11]. In Li *et al.* [11], PLS was used in cross-pose face recognition by constructing the relation between the coupled faces. In addition, PLS also has many extensions: In [21], bridge PLS was proposed by adding ridge-parameter in calculation to improve the efficiency in each iteration. Rosipal *et al.* [17] proposed the kernel PLS (KPLS) [17] by mapping the input variables into high dimension space so as to solve the nonlinear problem in linear space.

There is a common problem called semantic gap [15,25] existing in cross-modal retrieval. This problem is often solved by using label information due to its valuable semantic information [25]. In mobile devices, most users always give a tag to the microblogs or other things, such as the keywords given for each microblog, the items given by Amazon for each goods. The tags they provide equal label information in cross-modal retrieval. Then, by using the label information, the semantic gap decreases obviously. In Sharma's work [19], they proposed the framework named Generalized Multiview Analysis (GMA) to make use of labels for extracting multi-view features. Further, GMLDA and GMMFA, which are the application of GMA, shown competitive performances on the face recognition problem and cross-modal retrieval problem. In [10], Local Group based Consistent Feature Learning (LGCFL) was proposed which is a supervised joint feature learning method taking local group as priori.

Based on above discussion, we proposed a supervised algorithm, where the learned common feature space can be learned from two modalities. Our TPLS algorithm uses class indicator matrix indicating label information. At the same time, we introduce kernel partial least squares (KPLS) [17] to construct the relation between two multimedia modalities and the label modality. Because the KPLS can solve the nonlinear problem via linear method in its high dimension feature space, and it always gets better performance than PLS. In addition, we find that the common space constructed by label information are altered into two different spaces in KPLS iterative process, so we add novel constraint to minimize the divergence of label space. As a consequence of that, it makes the label space close to the other as possible as they can.

The remainder of the paper is organized as follows. In Sect. 2, we give a simple review of PLS and KPLS algorithms. Then, we will show our TPLS algorithm and its optimization in Sect. 3. In Sect. 4, experimental setting and result is shown. Finally, the conclusion is summarized in Sect. 5.

2 Preliminary

2.1 Partial Least Squares

PLS can construct the relation between two different modalities by maximizing the correlation between their latent variables. Let $X = [x_1, \cdots, x_n]^T$ represents the input variable with n training samples, where $x_i \in \mathbb{R}^{d1}$, $i = 1, \cdots, n$. Its latent variable represent as $V = [v_1, \cdots, v_n]^T \in \mathbb{R}^{n \times p}$, with $p \ll d1$. Respectively, let $Y = [y_1, \cdots, y_n]^T \in \mathbb{R}^{n \times d2}$ represent the output variable of training set. Its latent variables represents as $U = [u_1, \cdots, u_n]^T \in \mathbb{R}^{n \times p}$, with $p \ll d2$: PLS can be designed as:

$$\begin{cases} X = VP^T + \varepsilon_x \\ Y = UQ^T + \varepsilon_y \end{cases} \tag{1}$$

where the P and Q represent matrices of loadings, the ε_x and ε_y are the matrices of residuals. And by means of the low dimension latent variables V and U, we can further get a regression coefficient matrix $B \in \mathbb{R}^{d1 \times d2}$ to get the relation between X and Y:

$$B = X^T U (V^T X X^T U)^{-1} V^T Y \tag{2}$$

$$Y = X B^T + \varepsilon_B \tag{3}$$

where ε_B is the matrix of residuals.

PLS can be solved by traditional iterative algorithm calculating the first dominant eigenvector to get the weight vectors r, s. After i-th iteration, we can obtain the i-th latent vectors $v_i = Xr_i$, $u_i = Ys_i$ which respectively construct the i-dimension of latent variables V, U, and coefficient matrix B shown as Fig. 2. In traditional iterative algorithm, the object function can be described as follow:

$$[cov(v, u)]^2 = \max_{|r|=|s|=1} [cov(Xr, Ys)]^2 \tag{4}$$

Furthermore, according to the [16], we can also describe the object function of PLS as follow:

$$< v, u > = \arg\max_{r,s} < Xr, Ys > = \arg\max_{r,s} r^T X^T Y s \tag{5}$$

$$s.t. \quad r^T r = 1, s^T s = 1$$

where $< a, b > = a^T b$ is the inner product of vector.

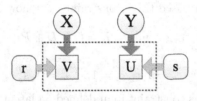

Fig. 2. The structure diagram of the PLS model.

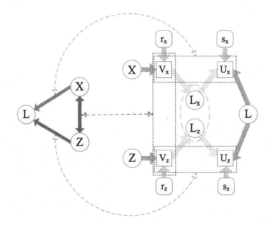

Fig. 3. The structure diagram of TPLS shows as above, in which the latent variables correlations constraint is indicated by three square dotted boxes, and Distance of target space constraint is indicated by the elliptical dotted boxes.

3   Trigonal Partial Least Squares

3.1   Trigonal Partial Least Squares Algorithm

Assume that there are two sets of multimedia kernel feature, $X = (x_1, x_2, \cdots, x_n)^T$, and $Z = (z_1, z_2, \cdots, z_n)^T$ from different modalities, where x_i is in n dimensions and z_i is also in n dimensions. Besides, the two sets of data are in c classes. And we construct the class indicator matrix $L = [l_1, l_2, \cdots, l_n]^T$, where l_i is a binary class indicator vector in c dimensions with all elements being zeros except for the place corresponding its classes. In other words, if sample z belongs to k-th class, then its class indictor vector l is $l_k = 1$ and $l_j = 0$ for $j \neq k$. As a result of two sets of data, we also have two class indicator matrices L_x, L_z respectively.

In our approach, we want to utilize label information to learn consistent feature representation. From Eq. 3, it is obvious that we need the regressor coefficient matrix B_x, B_y to map the input variables which are heterogenous features X, Z into the target feature spaces constructed by label information. So our approach learns four weight vectors r_x, s_x, r_z and s_z, which construct latent vectors $v_x = Xr_x$, $u_x = Xs_x$, $v_y = Vr_y$, $u_y = Vs_y$ respectively, and further construct the regressor coefficient matrices B_x, B_y via Eq. 2. And in our approach, we propose object function based three constraints as follow:

$$\arg \min_{r_x, r_z, s_x, s_z} F = G + D + P \tag{6}$$

$$s.t. \quad r_x^T r_x = 1, r_z^T r_z = 1, s_x^T s_x = 1, s_z^T s_z = 1$$

where G is a correlations constraint item defined on latent variables of heterogenous modal features and the class indicator matrix, D is a distance constraint item of iterative output to reduce the difference between two modalities in each iterations. P is a transformation constraint item, which is also a regularization item defined on the weight vectors. The TPLS model is demonstrated in Fig. 3.

Latent variables correlations constraint: Previous work [19] uses PLS in cross-modal retrieval by setting one modal feature as input variable, and the other modal feature as output variable. This usage of PLS constructs the relation between two sets of modal feature, via maximizing their corresponding latent variables. Actually, label information is available for training the embedding [6]. More specifically, many heterogenous modal pairs own a unique label which is variant from each other. Taking account that heterogenous modalities share common labels, if we can establish the relation between two modalities via label in more direct way, it will help to improve the performance of cross-modal retrieval. Based on this consideration, we take the label as the output variables. In our method, label information is represented by class indicator matrices L_x, L_z. And then the label information is introduced in our method through PLS model as follow,

$$X = V_x P_x^T + \varepsilon_x \tag{7}$$

$$L_x = U_x Q_x^T + \varepsilon_{Lx} \tag{8}$$

$$Z = V_z P_z^T + \varepsilon_z \tag{9}$$

$$L_z = U_z Q_z^T + \varepsilon_{Lz} \tag{10}$$

In the above equations, two PLS model are established: Eqs. 7 and 8 are one PLS model between X and its class indicator matrix L_x, Eqs. 9 and 10 are the other one PLS model between Z and L_z. Therefore, according to the Eq. 5, we get latent variables correlations constraint of multimedia and label as follow,

$$G_1 = \lambda_1 < Xr_x, Ls_x > + \lambda_2 < Zr_z, Ls_z > \tag{11}$$

Besides above two PLS processes, we want that X modal latent variables V_x and text latent variables V_z can express data variability to the utmost extent, like the PCA,

$$\begin{cases} var(v_x) \rightarrow \max \\ var(v_z) \rightarrow \max \end{cases} \tag{12}$$

at the same time, we also ask v_x to explain v_z as possible as it can, based on CCA, the correlation between v_x and v_y should be maximized as follow,

$$cov(v_x, v_z) = \sqrt{var(v_x)var(v_z)} r(v_x v_z) \rightarrow \max \tag{13}$$

based on Eqs. (12) and (13), we can obtain our multimedia-multimedia correlation constraint as follow,

$$G_2 = \lambda_3 < Xr_x, Zr_z > \tag{14}$$

thus, we put multimedia-label correlation constraint G_1 and multimedia-multimedia correlation constraint G_2 together to obtain latent variables correlations constraint G as follow,

$$G = G_1 + G_2 \tag{15}$$

Via Eq. 15, firstly, we can map both two multimedia modal features X, Z to a common feature space constructed by label information in two PLS processes. Secondly, we maximize the correlation between latent variables of two modalities based the idea of PCA and CCA, which is also a PLS process.

Distance of iterative output constraint: By using Eqs. 7–10, we construct two PLS processes which set the class indicator matrices L_x and L_z as output variables respectively. It is deserved to point out that, the initial class indicator matrices L_x, L_z are same, but as a result of different initial condition in each PLS iteration, caused by respective different latent vectors v_x and v_z, the class indicator matrices L_x and L_z will be different in the subsequent iterative process shown as Eq. 19. The reason for that result is that two PLS processes have different input variables which are multimedia modal feature respectively. Considering that we should achieve the consistent feature representation for different modalities, so the output latent space of U_x and U_z should be in the same space. However, looking for two output latent variables U_x and U_z in a same space is a too strong constraint, which will destroy the PLS training process. So we add a soft constraint which is the distance constraint between L_x and L_z in iterative process. That is, we want to make the distance of L_x and L_z close to each other in each iterative process, thus the initial condition of each iterative process is as proximal as possible, and finally the output latent spaces U_x and U_z are close to each other in PLS training process. According to above analysis, we use the distance of iterative output as follow,

$$D = \lambda_4 \| L_x - X r_x r_x^T X^T L_x - (L_z - Z r_z r_z^T Z^T L_z) \|_F^2 \tag{16}$$

By adding the D, we use the Frobenius norm of matrix to reach the initial condition in each iteration proximal to the each other in each iteration.

Transformation constraint: This constraint item can be expressed as follow to prevent over fitting:

$$P = \frac{1}{2}(\|r_x\|^2 + \|r_z\|^2 + \|s_x\|^2 + \|s_z\|^2) \tag{17}$$

3.2   RBF Kernel and Deflation Detail

Specially, we construct TPLS based on KPLS and adopt RBF kernel to extract the kernel feature X and Z, the RBF kernel function can be described as follow:

$$k(x, x') = \exp(-\frac{\|h - h'\|^2}{2l_{rbf}^2}) + \sigma_w^2 \varepsilon_{h,h'} \tag{18}$$

where $2l_{rbf}^2$, σ_w^2 denote the parameters of the RBF bandwidth and the variance of noise respectively.

As a result of using kernel features X, Z, we can obtain the deflation of the matrices X, Z, and class indicator matrices L_x, L_z. The deflation of KPLS is different to PLS after extraction of the i-th latent vector v as follow,

$$
\begin{cases}
X^{i+1} = X^i - v_x v_x^T X^i - X^i v_x v_x^T + v_x v_x^T X^i v_x v_x^T \\
Z^{i+1} = Z^i - v_z v_z^T Z^i - Z^i v_z v_z^T + v_z v_z^T Z^i v_z v_z^T \\
L_x^{i+1} = L_x^i - v_x v_x^T L_x^i \\
L_z^{i+1} = L_z^i - v_z v_z^T L_z^i
\end{cases}
\tag{19}
$$

3.3 Optimization

Then, we can obtain the optimal solution of TPLS by Stochastic Gradient Descent. And the gradient of each variable is solved as follow:

$$
\begin{cases}
\dfrac{\partial F}{\partial r_x} = \lambda_4 \dfrac{\partial D}{\partial r_x} - \lambda_1 X_z L s_x + \lambda_3 X_z Z r_z + r_x \\
\dfrac{\partial F}{\partial r_z} = \lambda_4 \dfrac{\partial D}{\partial r_z} - \lambda_2 Z_z L s_z + \lambda_3 Z_z X r_x + r_z \\
\dfrac{\partial F}{\partial s_x} = -\lambda_1 L X r_x + s_x \\
\dfrac{\partial F}{\partial s_z} = -\lambda_2 L Z r_z + s_z
\end{cases}
\tag{20}
$$

where

$$
\begin{cases}
\dfrac{\partial D}{\partial r_x} = -2(X^T A L_x^T X r_x + X^T L_x A^T X r_x) \\
+2(X^T L_x L_x^T X r_x r_x^T X^T X r_x + X^T X r_x r_x^T X^T L_x L_x^T X r_x) \\
\dfrac{\partial D}{\partial r_z} = -2(Z^T B L_z^T Z r_z + Z^T L_T A^T Z r_z) \\
+2(Z^T L_T L_z^T Z r_z r_z^T Z^T Z r_z + Z^T Z r_z r_z^T Z^T L_T L_z^T Z r_z) \\
A = L_x - L_z + Z r_T r_z^T Z^T L_z \\
B = L_x - L_z - X r_x r_x^T X^T L_z
\end{cases}
\tag{21}
$$

by using the SGD (Stochastic Gradient Descent), we only ask for a small part of training data in each iteration and then reduce the memory consumption in comparison with using BGD (Batch Gradient Desect) which makes usage of whole data set.

After we have solved the weight vectors r_x and r_z, we can further solved the latent vectors v_x and v_z. Similarly, we can get the latent vectors u_x and u_z. This was followed by solving the regression coefficient matrices B_x showing relation between between X and L and B_z showing relation between Z and L.

Algorithm 1. The algorithm of TPLS

Input:
  the image kernel feature X, the text kernel feature Z, the class indicator matrices L_x and
  L_z, the dimension of latent variables c, the batch setting for SGD β

Output:
  the weights matrices R^x and R^z, the latent variables V^x and V^z, the matrices of
  regression coefficients B^x and B^z, the residual matrices $\varepsilon_B^X, \varepsilon_B^Z$, and our regression
  models M^x, M^z

1: Initialize: $E_1^x = X$, $E_1^z = Z$, $F_1^x = L_x$, $F_1^z = L_z$
   for $k = 1$ to c **do**
2: Calculate the k-th weight vectors r_k^x, r_k^z by stochastic gradient descent algorithm using
   Eqs. 20 and 21
3: Calculate the k-th latent vector:
   $v_k^x = E_k^x r_k^x$, $v_k^z = E_k^z r_k^z$
4: Deflate E_k, F_k matrices using Eq. 19 respectively
   end
5: Calculate the regression coefficient matrices using Eq. 2 respectively
6: Calculate the residual matrix:
   $\varepsilon_B^x = F_k^x - E_k^x B^x$ $\varepsilon_B^z = F_k^z - E_k^z B^z$
7: Obtain the PLS regression models M^x, M^z using Eq. 3 respectively
8: **return** M^X, M^Z;

3.4   Computational Complexity Analysis

Lastly, we briefly analyze the computational complexity of TPLS method, which involves c iterations because of the c dimensions of latent variables, and each dimension is obtained by gradient descent algorithm, in which the max iterations set as z. Set n as the number of sample pairs in the training set, thus image feature and text feature are n dimensions as a result of RBF kernel. The computational complexity of TPLS is $O(czn^2)$.

3.5   Test TPLS on AVD

Above training process is done on a computer offline, after which, we test our model in an AVD, in which the configure setting is Device Nexus 5X API 26. And we just introduce how to test our TPLS on AVD in this section. The diagram of the how to apply TPLS in AVD is shown in Fig. 4.

As we have solved the B_x and B_z in the training process, we then store the two matrices and the feature of test set extracted in advance in the AVD. Then, according to the Fig. 4, our model can extract the image feature or the texts feature according to the query at first. The query feature is then project to the common space, which is the consistent representation in the picture. The parameters here are the B_x and B_z according to the query. After we have calculated consistent representation, we then get the similarity between query and respective dataset which is the feature of test set extracted in advance. This is followed by outputting the top ranked match results to the users. It is mentionable that we are not allowed to show the original images or texts, for these are too large to store in the mobile phone, such as the wiki dataset, which has only six hundred and ninety three images, requiring nearly 1 GB to store

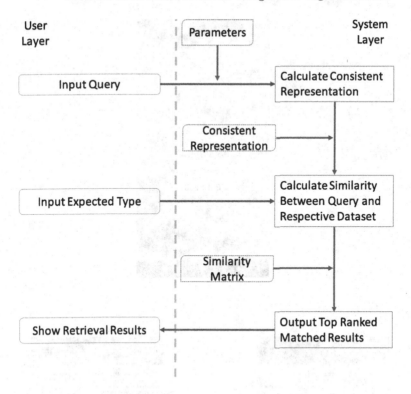

Fig. 4. The diagram of the how to apply TPLS in AVD.

them, let alone the texts. Hence, we just show the retrieval results in a form of their file name. Also, we show the demo of text-to-image task in the AVD in the Fig. 5, in which, we have two choices "IMAGE QUERY" and "TEXT QUERY" shown in the Fig. 5(a). After we click the button "TEXT QUERY", we step into the Fig. 5(b) which allows us to do the text-to-image task. Then, the texts have been keyed and we click the button "TEXT QUERY" again, the results are shown in the Fig. 5(c). Finally, we can click the triangle button to return the initial screen shown as the Fig. 5(a) which can do the new cross-modal retrieval.

4 Experiments

In this section, we test the proposed method on two popular databases to show its effectiveness.

4.1 Experimental Databases

Wiki is constructed from the Wikipedia including 2866 image-text pairs with 10 different categories. The images are represented by 128-dimensional vector based on SIFT descriptors and the texts are represented by 10-dimensional LDA

(a)

(b)

(c)

Fig. 5. The screenshots of the Demo in AVD in terms of the text-to-image task.

(latent Dirichlet allocation) feature. We randomly select 2173 image-text pairs for training set and 693 image-text pairs for testing set.

Flickr is a subset chosen from NUS-WIDE, which is crawled from the Flicker website. The Flickr database consists of 5730 single-label images associated with their tag text. For feature representation, we use 500-dimensional bag-of-words based on SITF descriptors as image feature and 1000-dimensional word frequency based tag feature as text feature. The image-text pairs are selected from NUS-WIDE in which have top-10 largest numbers of images. As a consequence of that, we randomly choose 2106 image-text pairs as training set and the rest 3624 pairs as testing set.

4.2   Evaluation Metric

In our experiment, we use MAP (Mean Average Precision) and PR (Precision-Recall) curve to show the effectiveness of MDLL.

MAP has been widely used to evaluate the overall performance of cross-modal retrieval, such as [10,15,20,24,28]. To compute MAP, we first evaluate the average precision (AP) of a retrieved database including N retrieved samples by $AP = \frac{1}{T} \sum_{r=1}^{N} E(r)\delta(r)$, where T is the number of the relevant samples in the retrieved database, $E(r)$ denotes the precision of the top r retrieved samples, and $\delta(r)$ is set to 1 if the r-th retrieved sample is relevant (on above three databases, a retrieved sample is relevant if it shares at least one label with the query) and $\delta(r)$ is 0 otherwise. Then by averaging the AP values over all the queries, MAP can be calculated.

Besides, PR curve is a classical measure of information retrieval or classified performance. Assume that the set S_1 includes the samples in which real labels are denoted by L_r. The classifier picks out the set S_2 samples in which labels are classified into L_r. In the set S_2, the samples in which real labels are L_r construct the set S_3. Thus, we can calculate the precision ratio: $PR = \frac{|S_3|}{|S_2|}$ and the recall ratio: $RR = \frac{|S_3|}{|S_1|}$, where $|A|$ means the number of elements in set A. Furthermore, we get different PR-RR values via the different classified setting and then draw precision-recall curve in which the vertical coordinate is precision ratio and the horizontal coordinate is recall ratio.

4.3   Compared Scheme

We compare our approach with PLS, Kernel PLS (KPLS), Semantic Correlation Matching (SCM), Correlated Semantic Representation (CSR), Generalized Mutiview Marginal Fisher Analysis (GMMFA) and Generalized Multiview LDA (GMLDA) in two retrieval tasks. In PLS, the modal latent variables are obtained by maximize the correlation between the latent variables of images and texts. KPLS maps the original feature into a high dimension feature space so as to construct the linear model to solve the nonlinear problem. As for SCM, it uses CCA to learn two maximally correlated subspaces, and then learns the logistic regressors in each subspaces. CSR learns a compatible function via structural SVM to get a joint image-text representation and an uniform formulation. GMMFA and GMLDA both use the framework Generalized Multiview Analysis.

4.4 Experimental Setting

In our experiments, we set the parameters for these two tasks as below: $\lambda_1 = \lambda_2 = 3$, $\lambda_3 = 1$, $\lambda_4 = 0.0001$. As for RBF Kernel, we set the RBF bandwidth l_{rbf} as 1 and the variance of noise σ_w^2 as 0 in data preparation for both two model. In addition, we set the dimension of latent variables $c = 200$ for TPLS on Wiki and $c = 300$ for TPLS on Flickr. With regard to the batch setting in SGD, we set $\beta = 100$ for TPLS on Wiki and $\beta = 150$ for TPLS on Flickr.

We use the classical precision-recall curve and the mean average precision (MAP) metric to evaluate the performance of algorithms.

4.5 Result on Wiki

Table 1. The MAP comparison results on Wikipedia database. The results shown in boldface are the best performance.

Methods	Tasks		
	im2txt	txt2im	Average
PLS [18]	0.207	0.192	0.199
KPLS [17]	0.260	0.201	0.231
SCM [15]	0.277	0.226	0.252
GMMFA [19]	0.264	0.231	0.248
GMLDA [19]	0.272	0.232	0.253
CSR [23]	0.243	0.201	0.222
TPLS	**0.312**	**0.241**	**0.277**

The MAPs of the different methods on the Wiki dataset are shown in Table 1. From Table 1, we can find the following scenes:

Firstly, the average MAP of KPLS outperform 16.1% than PLS, which indicates that mapping original feature into a high dimensional feature space via kernel function can get better performance. The advantage of KPLS motivates us to construct TPLS based on KPLS rather than PLS.

Secondly, for the supervised methods using the label information, such as SCM, GMMFA and GMLDA, they outperform the unsupervised algorithms PLS and KPLS by at least 7.36%. This indicates that the label information can provide the available information to improve the performance.

Finally, compared with KPLS, TPLS obtained 19.9% higher MAP, which validates the effectiveness of setting the label information as the output variables in TPLS. Compared with the supervised algorithm, the best performance of TPLS outperforms the second best SCM by 12.6% higer MAP in the image-to-text retrieval task. In the text-to-image retrieval task, TPLS outperforms the second best GMLDA by 3.9% higher MAP. All these results show the effectiveness of the constraints in TPLS.

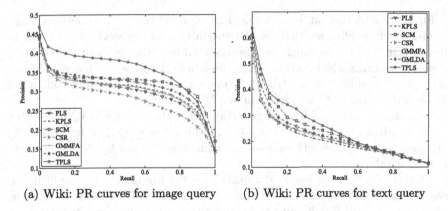

(a) Wiki: PR curves for image query   (b) Wiki: PR curves for text query

Fig. 6. Precision recall curves of cross-modal retrieval using both image and text queries on Wiki database.

In Fig. 6(a) and (b), we also show the PR curves of the different methods. We can know that TPLS performs against the other algorithms in both the retrieval tasks at the low levels of the recall. Considering that similar points need to be searched in a small neighborhood of the query, the low levels of recall are more practical in practice.

4.6 Result on Flickr

On the Flickr database, because the high dimension of the image and text features, we use PCA to reduce the dimension of the features. Especially, 95% information energy is preserved by the PCA. In Table 2, we show the MAP of the different methods on the Flickr database. From Table 2, it is easy to find the following scenes:

Table 2. The MAP comparison results on Flickr database. The results shown in boldface are the best performance.

Methods	Tasks		
	im2txt	txt2im	Average
PLS [18]	0.269	0.228	0.249
KPLS [17]	0.321	0.219	0.270
SCM [15]	0.215	0.136	0.176
GMMFA [19]	0.311	0.212	0.262
GMLDA [19]	0.299	0.188	0.244
CSR [23]	0.202	0.170	0.186
TPLS	**0.394**	**0.246**	**0.318**

172   J. He et al.

Firstly, KPLS gets at least 3.05% higher MAP than the supervised algorithms, which is different with the conclusion that the supervised methods are often better than the unsupervised methods. In fact, the text features of Flickr is the tag features which also includes the label information. That can further verify the availability of the label information in cross-modal retrieval, which has been indicated by the experiments on Wiki database.

Secondly, besides KPLS, it is remarkable that the average MAP of TPLS is 39.4%, which is 21.4% higher than the second best result (31.1% for GMMFA). This also verifies the effectiveness of our constraints shown by the experiments on Wiki database.

Thirdly, the texts in Filickr database have the tag information, which only uses several words rather than several paragraphs like Wiki databases. Under this condition, TPLS still gains the competitive results on the Flick database, which indicates that TPLS is also effective on the texts which is constructed just by several words.

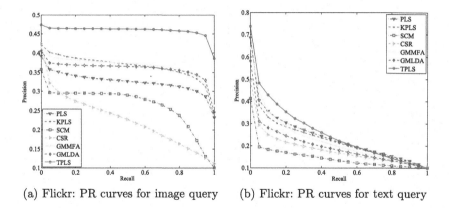

(a) Flickr: PR curves for image query   (b) Flickr: PR curves for text query

Fig. 7. Precision recall curves of cross-modal retrieval using both image and text queries on Flickr database.

Lastly, Fig. 7(a) and (b) show the precision-recall curves of the compared algorithms on two tasks. We can see that with the same recall rate, our approach reach higher precision than other algorithms at low level of recall, which is similar to the results on the Wiki database.

In summary, on both the Wikipedia and Flickr databases, TPLS can achieve the higher performance on both the retrieval tasks, which indicates the effectiveness of TPLS by introducing the label information and reducing the common space difference.

4.7   Running Time Analysis

In the case of the running time for each model, we also carry the experiments to compare it on the Wiki. The results are illustrated in Table 3. Then we can conclude as follow:

Firstly, TPLS solving by SGD is about 20 times faster than that by BGD. It shows the high efficiency of SGD through picking up a small part of training data rather than the whole data set.

Secondly, TPLS is time-consuming in training process. But the training is done offline and only once. Thus the training time cost is not as important as that of the testing time.

Thirdly, TPLS costs similar test time compared with other methods. The contributing factor accounting for the similar test time is that TPLS and other methods all learn a projector and makes test data comparable by multiplying the modal features and the projector directly.

Table 3. Calculational time on the Wiki database. The unit is second.

Methods	Tasks	
	Training time	Testing time
PLS [18]	26.92	17.69
KPLS [17]	84.41	17.76
SCM [15]	12.94	22.77
GMMFA [19]	184.12	16.96
GMLDA [19]	201.92	17.10
TPLS (BGD)	99606.11	16.76
TPLS (SGD)	4297.56	16.59

4.8   Exhibition of Retrieval Result

Besides above experiments, we also show the examples of queries and their results retrieved by RLPLS and GMLDA on the Wiki dataset shown as Fig. 8. The text query and the image of the ground truth are shown in the first column of the first and second row. The top five retrieved images of GMLDA and RLPLS are exhibited at the first and second row, respectively. The images with red frames are the wrong retrieval results based on their respective class. From the figure, we can know that all the retrieved images of RLPLS are correct while only the third retrieved image of GMLDA is related to the text query. At the third and forth row, we also show the image query, the text of the ground truth and the top five retrieved documents shown with their corresponding images of RLPLS and GMLDA. From the figure, we can also find that the third images of RLPLS is a wrong retrieval result.

Fig. 8. Two examples of queries and their results retrieved by RLPLS on the Wiki dataset. (Color figure online)

5   Conclusion

In this paper, we proposed a novel method to solve cross-modal retrieval problem under the hardware condition of the mobile devices, and apply it to the image-text cross-modal retrieval. In our approach, we regard the label information as the third modality so as to construct three KPLS between any two modalities. Furthermore, we add the distance constrain of target space so as to achieve learning the consistent feature representation in cross-modal retrieval. Experiments are carried out on two databases, Wikipedia and Flickr, showing that our proposed algorithm performs against existing competitive algorithms.

Later, we will look for more effective learning model to learn consistent representation. And we will design model to solve multi-label and cross-modal retrieval in mobile device.

Acknowledgement. This work was supported in part by the National Natural Science Foundation of China under Grant 61672497, Grant 61332016, Grant 61620106009, Grant 61650202 and Grant U1636214, in part by the National Basic Research Program of China (973 Program) under Grant 2015CB351802, and in part by the Key Research Program of Frontier Sciences of CAS under Grant QYZDJ-SSW-SYS013. This work was also partially supported by CAS Pioneer Hundred Talents Program by Dr. Qiang Qu.

References

1. Bai, S., Bai, X.: Sparse contextual activation for efficient visual re-ranking. IEEE Trans. Image Process. **25**(3), 1056–1069 (2016)
2. Bai, X., Bai, S., Zhu, Z., Latecki, L.: 3d shape matching via two layer coding. IEEE Trans. Pattern Anal. Mach. Intell. **37**(12), 2361–2373 (2015)
3. Chen, Y., Wang, L., Wang, W., Zhang, Z.: Continuum regression for cross-modal multimedia retrieval (ICIP 2012), pp. 1949–1952 (2012)

4. Deng, J., Du, L., Shen, Y.: Heterogeneous metric learning for cross-modal multi-media retrieval. In: International Conference on Web Information Systems Engineering, pp. 43–56 (2013)
5. Duan, L., Xu, D., Tsang, I.: Learning with augmented features for heterogeneous domain adaptation. arXiv preprint arXiv:1206.4660 (2012)
6. Gong, Y., Lazebnik, S.: Iterative quantization: a procrustean approach to learning binary codes. In: Proceedings of the IEEE Conference on Computer Vision and Pattern Recognition, pp. 817–824 (2011)
7. Hardoon, D.R., Szedmak, S., Shawe-Taylor, J.: Canonical correlation analysis: an overview with application to learning methods. Neural Comput. 16(12), 2639–2664 (2004)
8. He, R., Zhang, M., Wang, L., Ye, J., Yin, Q.: Cross-modal subspace learning via pairwise constraints. IEEE Trans. Image Process. 24(12), 5543–5556 (2015). A Publication of the IEEE Signal Processing Society
9. Jia, Y., Salzmann, M., Darrell, T.: Learning cross-modality similarity for multi-nomial data. In: Proceedings of the IEEE International Conference on Computer Vision, pp. 2407–2414 (2011)
10. Kang, C., Xiang, S., Liao, S., Xu, C., Pan, C.: Learning consistent feature representation for cross-modal multimedia retrieval. IEEE Trans. Multimed. 17(3), 370–381 (2015)
11. Li, A., Shan, S., Chen, X., Gao, W.: Cross-pose face recognition based on partial least squares. Pattern Recognit. Lett. 32(15), 1948–1955 (2011)
12. Lu, X., Wu, F., Tang, S., Zhang, Z., He, X., Zhuang, Y.: A low rank structural large margin method for cross-modal ranking, pp. 433–442 (2013)
13. Mao, X., Lin, B., Cai, D., He, X., Pei, J.: Parallel field alignment for cross media retrieval. In: Proceedings of the ACM International Conference on Multimedia, pp. 897–906 (2013)
14. Pereira, J.C., Coviello, E., Doyle, G., Rasiwasia, N., Lanckriet, G., Levy, R., Vasconcelos, N.: On the role of correlation and abstraction in cross-modal multimedia retrieval. IEEE Trans. Pattern Anal. Mach. Intell. 36(3), 521–535 (2014)
15. Rasiwasia, N., Pereira, J.C., Coviello, E., Doyle, G., Lanckriet, G.R.G., Levy, R., Vasconcelos, N.: A new approach to cross-modal multimedia retrieval. In: Proceedings of the ACM International Conference on Multimedia, pp. 251–260 (2010)
16. Rosipal, R., Krämer, N.: Overview and recent advances in partial least squares. In: Subspace, Latent Structure and Feature Selection, pp. 34–51 (2006)
17. Rosipal, R., Trejo, L.J.: Kernel partial least squares regression in reproducing kernel hilbert space. J. Mach. Learn. Res. 2, 97–123 (2002)
18. Sharma, A., Jacobs, D.W.: Bypassing synthesis: PLS for face recognition with pose, low-resolution and sketch. In: Proceedings of the IEEE Conference on Computer Vision and Pattern Recognition, pp. 593–600 (2011)
19. Sharma, A., Kumar, A., Daume III, H., Jacobs, D.W.: Generalized multiview analysis: a discriminative latent space. In: Proceedings of the IEEE Conference on Computer Vision and Pattern Recognition, pp. 2160–2167. IEEE (2012)
20. Song, G., Wang, S., Huang, Q., Tian, Q.: Similarity gaussian process latent variable model for multi-modal data analysis. In: Proceedings of the IEEE International Conference on Computer Vision, pp. 4050–4058 (2015)
21. Tang, J., Wang, H., Yan, Y.: Learning hough regression models via bridge partial least squares for object detection. Neurocomputing 152, 236–249 (2015)
22. Tenenbaum, J.B., Freeman, W.T.: Separating style and content with bilinear models. Neural Comput. 12(6), 1247–1283 (2000)

23. Verma, Y., Jawahar, C.: Im2text and text2im: associating images and texts for cross-modal retrieval. In: Proceedings of the British Machine Vision Conference (2014)
24. Viresh, R., Nikhil, R., Jawahar, C.V.: Multi-label cross-modal retrieval. In: Proceedings of the IEEE International Conference on Computer Vision, pp. 4094–4102 (2015)
25. Wang, J., Kumar, S., Chang, S.: Semi-supervised hashing for large-scale search. IEEE Trans. Pattern Anal. Mach. Intell. **34**(12), 2393–2406 (2012)
26. Wang, K., He, R., Wang, W., Wang, L., Tan, T.: Learning coupled feature spaces for cross-modal matching. In: Proceedings of the IEEE International Conference on Computer Vision, pp. 2088–2095 (2013)
27. Wang, S., Zhuang, F., Jiang, S., Huang, Q., Tian, Q.: Cluster-sensitive structured correlation analysis for web cross-modal retrieval. Neurocomputing **168**, 747–760 (2015)
28. Xie, L., Pan, P., Lu, Y.: A semantic model for cross-modal and multi-modal retrieval. In: Proceedings of the ACM Conference on International Conference on Multimedia Retrieval, pp. 175–182 (2013)
29. Yao, T., Kong, X., Fu, H., Tian, Q.: Semantic consistency hashing for cross-modal retrieval. Neurocomputing **193**, 250–259 (2016)
30. Yu, Z., Zhang, Y., Tang, S., Yang, Y., Tian, Q., Luo, J.: Cross-media hashing with kernel regression. In: IEEE International Conference on Multimedia and Expo, pp. 1–6 (2014)
31. Zhang, H., Liu, Y., Ma, Z.: Fusing inherent and external knowledge with nonlinear learning for cross-media retrieval. Neurocomputing **119**, 10–16 (2013)
32. Zhang, L., Ma, B., He, J., Li, G., Huang, Q., Tian, Q.: Adaptively unified semi-supervised learning for cross-modal retrieval. In: International Conference on Artificial Intelligence, pp. 3406–3412 (2017)
33. Zhang, L., Ma, B., Li, G., Huang, Q., Tian, Q.: Pl-ranking: a novel ranking method for cross-modal retrieval. In: Proceedings of the ACM International Conference on Multimedia, pp. 1355–1364 (2016)
34. Zhang, L., Ma, B., Li, G., Huang, Q., Tian, Q.: Cross-modal retrieval using multi-ordered discriminative structured subspace learning. IEEE Trans. Multimed. **19**(6), 1220–1233 (2017)
35. Zhuang, Y., Wang, Y., Wu, F., Zhang, Y., Lu, W.: Supervised coupled dictionary learning with group structures for multi-modal retrieval. In: AAAI Conference on Artificial Intelligence (2013)

Author Index

Printed in the United States
by Bookmasters

Printed in the United States
By Bookmasters